Building Pathology
Principles and Practice
Second Edition

David S. Watt
BSc (Hons), DipArchCons (Leic), PhD, MSc, FRICS, IHBC
Hutton + Rostron Environmental Investigations Limited

Blackwell
Publishing

Blackwell Publishing editorial offices:
Blackwell Publishing Ltd, 9600 Garsington Road, Oxford OX4 2DQ, UK
　Tel: +44 (0)1865 776868
Blackwell Publishing Inc., 350 Main Street, Malden, MA 02148-5020, USA
　Tel: +1 781 388 8250
Blackwell Publishing Asia Pty Ltd, 550 Swanston Street, Carlton, Victoria 3053, Australia
　Tel: +61 (0)3 8359 1011

First edition published by Blackwell Science Ltd, a Blackwell Publishing company 1999
Second edition published by Blackwell Publishing Ltd 2007

2　　2008

ISBN: 978-1-4051-6103-9

Library of Congress Cataloging-in-Publication Data

Watt, David, 1963-
　Building pathology: principles and practice / David S. Watt. – 2nd ed.
　　p. cm.
　Includes bibliographical references and index.
　ISBN 978-1-4051-6103-9 (pbk. : alk. paper)
　1. Building failures–Investigation.　2. Buildings–Repair and reconstruction.
　3. Buildings–Defects.　I. Title.

TH441.W38 2007
690'.21–dc22　　　　　　　　　　　　　　　　　2007009090

A catalogue record for this title is available from the British Library

Set in 10/13pt Palatino
by Aptara, India

Contents

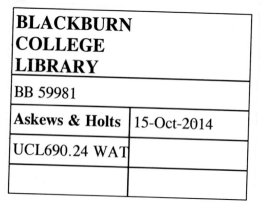

Preface to Second Edition

Building pathology, both as a term and as an overall concept, is becoming more widely used to define a holistic approach to understanding buildings. Such an approach requires a detailed knowledge of how buildings are constructed, used, occupied and maintained, and the various mechanisms by which their structural, material and environmental conditions can be affected. It is, by necessity, an interdisciplinary approach and requires a wider recognition of the ways in which buildings and people respond and react to each other.

The purpose of this book is to introduce the concept of building pathology and, with it, bridge the gap between current approaches to the surveying of buildings and the detailed study of defect diagnosis, prognosis and remediation. It has been written as a textbook for practitioners and students of built environment disciplines, and will hopefully be of use to others who are responsible for managing buildings and their sites.

A common criticism of books concerned with the survey, repair and maintenance of buildings is that no absolute answers are given, whether for the diagnosis of defects or specification of repairs. The reason is that such answers usually require more information than can readily be given in the pages of a book. It is partly in response to this omission that the present text seeks a greater awareness and comprehension of buildings and their users to assist in the design and implementation of specific and appropriate remedial measures.

Since the first edition of this book in 1999 the term 'building pathology' has become more widely accepted, and is used both by professional offices and academic institutions. The emphasis remains with understanding the underlying basis on which buildings are designed, constructed and utilised, and how they might sensibly and economically be managed, repaired and maintained now and in the future. This is set in the context of significant change in how we build and use such buildings. Environmental awareness and the drive toward sustainability are moving us in new directions, whilst the information revolution brought about by the internet and instant communication has altered everyday patterns of work and leisure. There are broadening skills gaps in the construction industry and a declining knowledge base from which we can draw, at a time when modern

forms of construction require a set of skills different from those needed only a generation ago. It is a time of change for us and our buildings.

The author is grateful to the following people for help, advice and information given during the preparation of this book:

George Ballard, GBG, Swaffham Bulbeck, Cambridgeshire
Professor John Burland, Imperial College, London
Professor May Cassar, Centre for Sustainable Heritage, University College London
Bob Kindred MBE, Ipswich Borough Council
Dr Maria Kliafa, National Gallery of Athens, Greece
Dr Gerard Lynch, Woburn Sands, Buckinghamshire
Bill Martin, English Heritage, London
Gerallt Nash, St Fagans: National History Museum, Cardiff
Neil Noble, Practical Action, Bourton-on-Dunsmore, Warwickshire
Richard Pollock, Burnett Pollock Associates, Edinburgh

Particular thanks are due to my wife, Dr Belinda Colston, for her help in proofreading and preparation of illustrations.

Unless otherwise stated, all photographs are by the author.

David S. Watt
January 2007

Chapter 1

Introduction

What is building pathology?

The term *pathology* is defined as the systematic study of diseases with the aim of understanding their causes, symptoms and treatment. In a medical context, the person becomes the subject of detailed examination and investigation, with consideration given to age, health and lifestyle. A similar approach is relevant in the study of buildings, and it is this methodical and often forensic practice that has come to be termed *building pathology*.

Building pathology, both as a term and as an overall concept, is becoming more widely used to define the holistic approach to understanding buildings. Such an approach requires a detailed knowledge of how buildings are designed, constructed, used and changed, and the various mechanisms by which their material and environmental conditions can be affected. It is, by necessity, an interdisciplinary approach and requires a wider recognition of the ways in which buildings and people respond and react to each other.

The definition of building pathology given by the Association d'Experts Européens du Bâtiment et de la Construction (AEEBC, 1994) draws attention to three separate, though interrelated, areas of concern:

- identification, investigation and diagnosis of defects in existing buildings;
- prognosis of defects diagnosed, and recommendations for the most appropriate course of action having regard to the building, its future and resources available; and
- design, specification, implementation and supervision of appropriate programmes of remedial works; monitoring and evaluation of remedial works in terms of their functional, technical and economic performance in use.

Other definitions include:

- the study of failures in the interrelationship of building structures and materials with their environments, occupants and contents (Hutton & Rostron, 1989);
- the study of failures over time in building materials and components (Groák, 1992, p. 105);
- the systematic treatment of building defects, their causes, their consequences and their remedies (CIB W86 Building Pathology Commission, 1993); and
- the scientific study of abnormalities in the structure and functioning of the building envelope and its parts; it seeks to study the interrelationships of building materials, construction, services and spatial arrangement with their environments, occupants and contents (Singh, 1997).

The CIB (International Council for Research and Innovation in Building and Construction, formerly International Council for Building) was established in 1993 to stimulate and facilitate international cooperation and information exchange. It has since developed into a world-wide network whose members are active in over 50 Commissions covering all fields in building and construction-related research and innovation. The objectives of the Building Pathology Commission (W86) are to produce information that will assist in the diagnosis and prevention of significant defects and failures in the design, construction and use of buildings, and to consider technical aspects of defects and failures in buildings of all types.

Although each definition places a slightly different emphasis on the nature and extent of the discipline, it is clear that building pathology, in its widest sense, is concerned principally with defects and associated remedial action. The purpose of this book is therefore to expand the range of investigation normally undertaken in the surveying of buildings, and to draw together various categories of information that are required to make informed decisions about how such buildings might be repaired, maintained and best utilised now and in the future.

Why take a holistic approach to understanding buildings?

Buildings do not exist in isolation, but instead represent various levels of action and interaction between Man and his surroundings – on the one hand they can be expressions of creative impulse, and, on the other, simple statements of functional need. In whichever form the building exists, it is a physical response to people, place and the environment (Fig. 1.1). Shifts in the balance between these three factors are responsible for many

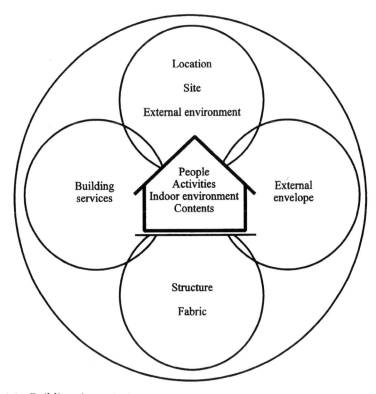

Fig. 1.1   Buildings in context.

of the decisions around which buildings are built, occupied, adapted and ultimately destroyed.

In order to understand a building, it must first be considered in context, from when it was designed and built, through changes over time, to its present use today. This progression takes into account various actions, some significant and others more mundane, but all giving information that may have relevance to understanding the building in the context of the present. Such an approach has much in common with archaeology, combining aspects of discovery, scientific analysis and creative imagination, but with the wider objective of informing decisions that will affect the present and the future (Fig. 1.2).

Taking a wider view of a building thus requires a level of understanding that, apart from simple examples, will often require the knowledge and experience of various disciplines. Those who might commonly be called upon to offer advice or an opinion as part of an interdisciplinary team may include:

- administrators and facilities managers
- archaeologists

Fig. 1.2   Archaeological investigation helping to understand the history and development of buildings.

- architects and designers
- art and architectural historians
- building services engineers
- building surveyors
- ecologists
- environmental and material scientists
- garden and landscape consultants
- general practice, planning and development surveyors
- interior designers
- quantity surveyors
- structural and civil engineers

Additional advice, information or comment may also be received from:

- amenity societies and pressure groups
- governmental departments and organisations
- non-governmental organisations
- owners and occupiers
- public utilities
- service staff (e.g. caretakers, cleaners, ground staff, security)
- statutory authorities:

- building control
- conservation
- countryside
- environmental health
- planning
- transportation

Further, more specialised, information may be required from other groups or individuals when dealing with specific building types or situations. Those who might contribute, for instance, to an understanding of historic buildings or ancient monuments, such as conservators, curators and craft-workers, have been considered in detail in a series of 'profiles' drawn up to demonstrate the interdisciplinary nature of conservation projects (COTAC, 1994), whilst other sources of information may have to be sought and examined for each specific need.

Such needs as are required to form an understanding of a building must consider the building in context with its location and use. Consideration of one without the others is a common fault that may ultimately lead to dissatisfaction, unnecessary expense or unjustifiable change.

Needs of the building

Buildings, together with their contents, present a complex assembly of materials and parts. Each material, whether it forms an identifiable element or component of the construction or part of the internal fabric, has its own characteristics and requirements. Traditional buildings, which are essentially a collection of natural materials, rely on soft mortars, sacrificial renders, moisture-vapour permeable plaster and finishes, and natural ventilation to retain their integrity and cohesion. This is in contrast to more modern buildings that make use of cements and concretes, plastics, composites, and other artificial or man-made materials to fulfil the requirements of client, designer and statutory authority (Fig. 1.3).

The needs of the building, of whatever age or construction, must be understood, respected and responded to if it is to function to an acceptable standard. It is these needs, and the question of what is 'acceptable', that will be considered in later parts of this book.

Needs of the building user

The use and function of buildings change over time, and with each change comes a different, and often conflicting, set of requirements. These user

Fig. 1.3   The design, construction and use of today's buildings differ in many ways from those of previous generations.

requirements will typically leave evidence in the form of physical changes to the structure, fabric and services of the building; personal recollections and remembrances; and associated documentation. Each of these levels of evidence will provide potentially useful information to be collected and considered when attempting to understand a building or collection of buildings.

The relevance of building pathology

The relevance of building pathology to practitioners and students of built environment disciplines, and others who are responsible for managing buildings and their sites, lies principally in the need for more accurate and

appropriate information on which to base decisions. This need may arise for a variety of reasons:

- determine financial security against an intended loan or mortgage, or change of ownership;
- provide confidence for a potential purchaser or tenant undertaking repair liabilities, either by way of a report commissioned directly by the purchaser or by the vendor wishing to confirm or disclose material facts;
- determine stability and risk of failure following natural or man-made disasters;
- establish liability for disrepair (dilapidations);
- diagnose defects when symptoms appear to occupiers;
- determine the effectiveness of past repairs or maintenance;
- assess levels of disrepair in advance of legal proceedings;
- ensure compliance with legal requirements;
- understand and document factors affecting condition;
- provide a basis for planned work (repair, maintenance); or
- provide a basis for physical change (adaptation, change of use).

Whatever the reason, this need for accurate and appropriate information, acquired at a cost that is acceptable to the client, will require a change in the ways in which buildings are perceived and dealt with. The acceptance and practice of building pathology, providing a holistic approach to understanding buildings, will add an extra dimension to what many professional advisers are already able to offer. As such, its relevance needs to be acknowledged and understood, and its principles adopted.

The principles and practice of building pathology

The principles upon which building pathology is based rely on a detailed knowledge of how a building is designed, constructed, used and changed, and the various mechanisms by which its structural, material and environmental conditions can be affected. It is more than just a detailed building survey, for it acknowledges the relative importance of both people and place. Such a comprehensive approach to understanding buildings offers potential for developing a deeper understanding and providing more useful information.

The following chapters are laid out to provide a logical progression, with consideration of buildings; building performance; causes and effects of defects, damage and decay; survey and assessment; remediation in practice; and principles of building management and aftercare.

References

AEEBC (1994) *Academic Guidelines: Policy Regarding Degree Validation*, London and Brussels: Association d'Experts Européens du Bâtiment et de la Construction.

CIB W86 Building Pathology (1993) *Introduction*. CIB Report 155, June.

COTAC (1994) *Multi-Disciplinary Collaboration in Conservation Projects in the UK*. 13 July. London: Conference on Training in Architectural Conservation.

Groák, S. (1992) *The Idea of Building: Thought and Action in the Design and Production of Buildings*. London: E. & F.N. Spon.

Hutton + Rostron (1989) *Building Pathology Conference (BP89)*. Gomshall: Hutton + Rostron.

Singh, J. (1997) *Historic Building Pathology and Health*. Paper presented at The Health of our Heritage conference, 2nd RIBA National Conservation Conference, 9 May, Bath.

Chapter 2
Understanding Buildings

What is a building?

What is a building? Although this might at first sight appear to be a relatively straightforward question, it can nevertheless be answered in a variety of ways. To most people who live and work in buildings, they are merely containers for activities that require shelter from the external environment. Such containers may vary in complexity from simple bus shelters to elaborate cathedrals, or from traditional forms of construction to those that rely on sophisticated building services in order to create and maintain specified environmental conditions.

The image that a building acts as a container or envelope, which buffers or filters external conditions for internal needs, is one that is widely used in understanding how buildings work. An analogy of a building acting like a skin, which surrounds the occupants and modifies environmental conditions, is similarly useful in that it indicates how it must be strong, resilient and able to adapt to changing conditions if it is to succeed and survive. Self-healing, both as a natural ability of the skin and inspiration for futuristic surface materials, takes this concept still further.

This image of a building behaving as a skin has been advanced with the notion that there is no such thing as a building at all! Instead, there is a series of layers or boundaries – *shell, services, scenery* and *set* proposed by Duffy (1990, p. 17) and *site, structure, skin, services, space plan* and *stuff* proposed by Brand (1994, p. 13) – which tear or shear due to different rates of change. This is again useful in emphasising that buildings are more than just bricks and mortar (Fig. 2.1).

Whether buildings are *more* than the sum of their component parts, representing a synergistic relationship between building and user, is a matter for personal opinion and debate. This is not to suggest the existence of a built 'superorganism', similar to that of James Lovelock's *Gaia* (Lovelock, 1995), but the concept that a building has a birth, life and death is, however, known in various parts of the world, and is acknowledged in the temple building of the ancient Mayan civilisation of central America.

9

Fig. 2.1 Buildings may be seen as a collection of different layers that react and respond to one another, but ultimately have to fit together, as with this Russian doll, in order to work. (Photograph by John Stanley.)

The success of a building in fulfilling its basic duties of containment and shelter depends on a series of related and interrelated issues. Much has been written on design theory and practice, and whilst it is not the purpose of this book to comment on how new buildings are procured, designed and built, it is useful to consider five opinions on the subject:

'In *Architecture* as in all other *Operative* Arts, the *end* must direct the *Operation*. The *end* is to build well. Well building hath three *Conditions*. *Commodity* [user satisfaction], *Firmness* [structurally sound], and *Delight* [aesthetically pleasing].' (Sir Henry Wotton, 1624)

'We start with the ground. This is rock and humus. A building is planted to survive the elements – the ground already has form. Why not begin by accepting that? Is the ground a prairie, square, flat? Is the ground sunny or the shaded slope of some hill, high or low, bare or wooded, triangular or square? Has the site features, trees, rocks, streams or a visible trend of some kind? Has it some fault or a special virtue or several? So essentially the site is the starting point of design.' (Frank Lloyd Wright, n.d.)

'Man puts available materials together to form shelter in such a way as to modify the indigenous climate in order to provide a satisfactory climate of comfort and convenience within. If the climate concept includes the cultural, social, political, aesthetic climates in addition to the physical one it suggests

three kinds of information are needed [pattern of activities, available site with its indigenous climate and building technology]. Without satisfaction, an individual may be unhappy, inefficient and uncomfortable.' (Geoffrey Broadbent, 1973)

'The reason for architecture is to encourage ... people ... to behave, mentally and physically, in ways they had previously thought impossible.' (Cedric Price, 1975)

'What should we ask of a new shelter? We should ask for protection from the elements, an adequate level of comfort and a pleasurable environment that enhances our life. These features should be supplied economically, simply, reliably. Shelter should not dominate our lives but rather make minimum impact upon us. Ideally, a shelter should make us aware of the beauties and delights of nature rather than remove us from them.' (Rodale, 1982)

Buildings are also expressions of the people and society that built them – this forms part of the national identity (Fig. 2.2). Changes in society are thus reflected in how and when buildings are designed, constructed, utilised, adapted and ultimately destroyed. Some of the most important concerns to have shifted building design and construction throughout history have been those of comfort and security – each has forced change that today represents history, whether it be architectural or social, political or economic.

Fig. 2.2 The Taj Mahal at Agra, India was built as a lasting symbol of love by one man for his wife. Today, its recognition is universal.

These changes, and the changing expectations of comfort standards, continue to this day with time- and labour-saving devices and gadgets, greater efficiency and control over heating, greater delineation of private space, intruder and fire detection systems, advanced electronic and telecommunications, home entertainment, and so on. Such changes, and corresponding shifts in attitude, will continue for as long as there is freedom of choice and action.

Although change is evident in how we use our buildings, there are nevertheless reminders of the past embedded in the buildings of the present. Such symbols of fashion or sentiment represent a visible link to earlier principles and practices, albeit often misunderstood and misapplied. This preoccupation with the past, fuelled by a growing interest and awareness in the nation's heritage, also provides a tension between old and new, witnessed by buildings that are out of harmony with their surroundings in place and time.

There is also a demand for 'green' and sustainable buildings, which have a minimal effect on the natural environment, and yet provide all the comforts and security that twenty-first century technology and science can provide (see Chapters 6 and 7). This duality provides the challenge that will ultimately take buildings and their construction into the next decades.

Perceptions of buildings

Since people first began to think of buildings as commonplace (probably with the advent of mass housing in the twentieth century), rather than essential for their survival, our perception of and regard for the built environment has progressively diminished. Buildings might thus be many things to many people, yet for much of the time their presence and purpose are ignored.

Whether one likes a building or not depends on personal preference and refinement. This is derived from a host of conscious and subconscious judgements, including personal values, beliefs and meanings (Fig. 2.3); knowledge and experience of a building or space; and mental or visual stimuli based on prompts such as books, films and childhood memories. These personal, and often intimate, perceptions or sensations may include:

- light and dark
- hot and cold
- dry and humid
- sunshine and shadow
- colour and texture

Fig. 2.3   Rushton Triangular Lodge, Northamptonshire. In plan, this building is an equilateral triangle, with three storeys having three windows on each side and on each floor. Each side has three gables, rising to three tapering pinnacles. At the intersection of the roof is a three-sided chimney stack. Below the gables is a frieze with a continuous inscription carried round the three sides, each side (33 ft long) bearing 33 letters. The Lodge, built by Sir Thomas Tresham in 1593–97, is symbolic of the Holy Trinity and linked to the doctrine of the Mass, and contains allusion to both religious literature and personal imagery (Isham, 1995).

- smells and odours (e.g. musty cellar)
- sound and silence (e.g. music)
- location and situation
- size and scale
- context and use
- character and association (e.g. 'haunted house')
- people and contents

As well as such manifest observations, buildings, as with pictures and sculptures, are able to cause the user or observer to experience their

surroundings in less apparent ways. 'Feelings' or sensory responses that might be experienced when in and around buildings may thus indicate a latent awareness of what is 'good' and 'bad'. These feelings have been used by designers throughout history to bring about differing emotions, sensations or behaviour that reflect the nature and use of the building. Such stimulation or arousal is one of the essential elements of good architecture (Table 2.1).

Table 2.1 Various feelings generated by architecture.

'Good' feelings		'Bad' feelings	
• homely	• welcoming	• claustrophobic	• lonely
• peaceful	• comfortable	• intimidating	• morbid
• spiritual	• spacious	• overwhelming	• isolated
• restful	• uplifting	• demoralising	• uncomfortable
• atmospheric	• exciting	• cramped	• impoverished
• inspiring	• breathtaking	• oppressive	• squalid

These judgements are, however, essentially subjective in nature, and may only partly answer the question of whether a building is really 'good' or 'bad'. Objectivity comes from acknowledging the various requirements of the building and assessing it against various accepted criteria. These might include:

- fitness for purpose (e.g. needs and expectations)
- accessibility (e.g. able and disabled persons)
- energy efficiency (e.g. thermal insulation, carbon dioxide emissions, SAP energy assessment ratings)
- sustainability (e.g. resource management)
- condition (e.g. repair and maintenance)
- performance in use (e.g. life cycle costs, levels of obsolescence)

Perceiving or 'seeing' buildings for what they are, as well as what they have been and might become, demands consideration at various levels. Most people can understand buildings in terms of construction, space and cost, usually based on their own experiences of buying and selling, yet each of these considerations, and more, can form the basis for detailed inquiry that takes the object of everyday life into the realms of academic study. The depth of such investigation will depend on the reasons for wishing to 'see' and understand the building – in the case of historic buildings, this may include learning the 'language' of the architecture in order to 'read' the design.

Classification of buildings

A simple way to see and understand buildings is to classify them according to how they look and what they do. Such classification typically attempts to bring together a number of similar building types or uses for one or more reasons. At a general level, buildings may be seen in terms of their history and, as such, can be categorised according to their age, stylistic influences and manner of construction. This categorisation forms the basis for the system of listing for buildings of architectural or historic interest (Department of the Environment/Department of National Heritage, 1994, PPG 15, Section 6.11):

- age and rarity are relevant considerations, particularly where buildings are proposed for listing on the strength of their historic interest. The older a building is, and the fewer the surviving examples of its kind, the more likely it is to have historic importance;
- all buildings built before 1700 that survive in anything like their original condition are listed;
- most buildings of about 1700 to 1840 are listed, though some selection is necessary;
- after about 1840, because of the greatly increased number of buildings erected and the much larger numbers that have survived, greater selection is necessary to identify the best examples of particular building types, and only buildings of definite quality and character are listed;
- for the same reasons, only selected buildings from the period after 1914 are normally listed;
- buildings that are less than 30 years old are normally listed only if they are of outstanding quality and under threat; and
- buildings that are less than ten years old are not listed.

Beyond this, classification is sought by legislation or regulation in order to impose order and conformity for a particular purpose. Such purposes are typically to do with land use, planning, and matters of health and safety:

- The Town and Country Planning (Use Classes) Order 1987, as amended, prescribes 13 classes of use within which change can take place without constituting development and so requiring planning permission. These classes of use are: A1 Shops; A2 Financial and professional services; A3 Restaurants and cafés; A4 Drinking establishments; A5 Hot food takeaways; B1 Business; B2 General industrial; B8 Storage or distribution;

C1 Hotels; C2 Residential institutions; C3 Dwellinghouses; D1 Non-residential institutions; and D2 Assembly and leisure.
- Approved Document B of the Building Regulations 2000 (as amended), which is concerned with fire safety, identifies seven purpose groups (Table D1) that can refer either to a whole building or to particular compartments within a building: residential (dwellings); residential (institutional); office; shop and commercial; assembly and recreation; industrial; and storage and other non-residential.
- The Fire Precautions Act 1971 is concerned with '...the protection of persons from fire risks, and for purposes connected therewith'. The Act stipulates that a 'fire certificate', which is issued by the fire authority, is required for all premises identified by 'designating orders'. These currently include: hotels and boarding houses; factories, offices, shops and railway premises; buildings containing two or more factory, office, shop or railway premises; and factory premises. The 1971 Act has been repealed by the Regulatory Reform (Fire Safety) Order 2005, and fire certificates will no longer be used.

Although such classifications allow buildings and building types to be 'understood' at the most general level, they do not attempt to distinguish or define one particular building from another. For this, it is necessary to understand the requirements of the individual buildings, and the expectations of those who own or use them.

Requirements of buildings

In order to be successful, the design and construction of a building has to consider a variety of issues. Or, to put this another way, a building, once it has been built, must fulfil certain criteria. These may be considered as being:

- functional requirements
- performance requirements
- statutory requirements
- user requirements

Functional requirements

Every building, regardless of its original, intermediate or ultimate use, can be expected to fulfil certain basic functional requirements. These requirements are primarily concerned with protection from the external

environment, human comfort, and organisation of activity and space (Fig. 2.4). Other functional needs might include the creation of a particular sense of identity or place, and the control of competing or conflicting internal uses.

Unless the function of a building is known, it cannot be judged to be good or bad.

Fig. 2.4 Comlongon Castle, Dumfries, Scotland. The massive fourteenth-century tower house, built at a time when defence was a principal functional requirement in the border region of Scotland, compares to the late nineteenth-century mansion house designed largely for appearance and comfort.

Performance requirements

For a building to be successful, it must satisfy the basic functional requirements noted above. The way in which it meets these demands, both as a building and as a collection of related and interrelated parts, may be determined by how it performs in relation to a number of defined performance measures or standards.

The performance requirements of a building and its various elements may be considered under the following headings, as illustrated in Fig. 2.5:

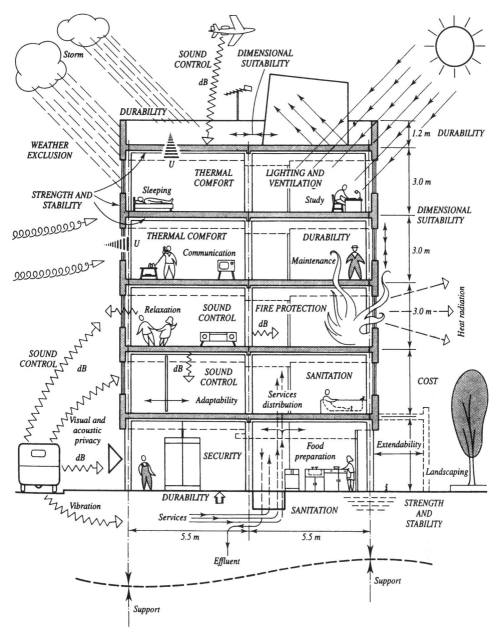

Fig. 2.5 Performance requirements for buildings. Reproduced from *Mitchell's Introduction to Building* by D. Osbourn and R. Greeno (1997). (Reprinted by permission of Addison Wesley Longman Limited.)

- access and egress
- appearance
- durability
- dimensional stability
- strength and stability
- weather exclusion
- sound control
- thermal comfort
- fire protection
- lighting and ventilation
- sanitation
- security
- cost

Many of these performance requirements form the basis of statutory and non-statutory demands that need to be met, in relation to both new buildings and the continued use of those already in existence.

Statutory requirements

There are various statutory and non-statutory requirements that make demands on those who design, build, manage, repair, maintain, occupy or demolish buildings. In practice, many of these demands are made in relation to the health, safety and well-being of such persons.

Some of the principal sources for statutory and non-statutory requirements (many with subsequent amendments and revisions) are:

- London Building Acts 1930–39
- Public Health Acts 1936, 1961
- Factories Act 1961
- Offices, Shops and Railway Premises Act 1963
- Fire Precautions Act 1971
- Health and Safety at Work, etc Act 1974
- Ancient Monuments and Archaeological Areas Act 1979
- Building Act 1984
- Housing Act 1985
- Town and Country Planning Act 1990
- Planning (Listed Buildings and Conservation Areas) Act 1990
- Environmental Protection Act 1990
- Planning Policy Guidance 16: Archaeology and Planning 1990
- Workplace (Health, Safety and Welfare) Regulations 1992
- Control of Substances Hazardous to Health Regulations 1994

- Construction (Design and Management) Regulations 1994
- Planning Policy Guidance 15: Planning and the Environment 1994
- Disability Discrimination Act 1995
- Reporting of Injuries, Diseases and Dangerous Occurrences Regulations 1995
- Construction (Health, Safety and Welfare) Regulations 1996
- Party Wall etc Act 1996
- Fire Precautions (Workplace) Regulations 1997
- Management of Health and Safety at Work Regulations 1999
- Building Regulations 2000
- Control of Asbestos at Work Regulations 2002

User requirements

The user of a building can expect to live or work in a space that satisfies basic human requirements and, in addition, certain needs that are specific to the activities being performed. The ways in which these are met, and whether one is in conflict with the other, is a measure of how appropriate the building is for the activity or activities in question. Fitness for purpose is thus an important measure of how a building matches the requirements of its user.

User requirement studies attempt to identify purpose in terms of activities (the things people do) and human needs (physical, psychological, physiological and social), and for a building to be fit for its purpose it must allow its occupants to carry out their activities economically and conveniently, and have a satisfactory environment to suit the user (Fig. 2.6). Such a study will typically consider:

- classification of user (e.g. task orientation)
- analysis of activities (e.g. social interaction)
- requirements of space (e.g. circulation in and around building)
- environmental conditions (e.g. sensory stimulation)
- structural implications (e.g. compatibility)
- cost (e.g. improvements)

Where buildings are designed or adapted for specific needs, these basic requirements may be replaced or supplemented by further considerations. These may be prescriptive in nature and, as such, might include requirements under specific headings (such as floor loadings or lighting levels). In such a situation, study of the particular needs of the user will assist in identifying what the building has to provide in order to satisfy user activities and human necessity (see Chapter 6).

Fig. 2.6 The activities and requirements of a building user determine the form of initial construction and subsequent alteration. The three-dimensional qualities of a building and the interconnectedness of its rooms and spaces may be appreciated in this doll's house, built in *c.* 1735 and housed at Nostell Priory in West Yorkshire. (Photograph reproduced courtesy of the National Trust Photographic Library. Copyright © NTPL/Mark Fiennes.)

The user requirements within a particular building may at times conflict with the structural, material and/or environmental needs of the building or its contents. This may be of particular concern when dealing with historic buildings, where careless alterations or adaptations can cause irreparable damage both to the structure and fabric of the building, and to the aesthetic qualities of its spaces. Where such conflict exists, it is important that the needs of both the building and its user(s) are clearly recognised, the significance of the building assessed, and the implications of bias or compromise fully understood.

Our expectations of buildings

Our understanding of buildings is based largely on expectation. When required to consider what buildings are, what functions they perform and what faults they might have, much depends on what we anticipate in their design, construction and usage. Some of this information is inherent within the building itself, yet much requires investigation to ascertain detail and fact.

The 'use' and 'type' of a particular building are often evident in how it looks and performs:

- agriculture – barn
- commerce – shop, office
- defence – castle
- education – school
- entertainment – cinema, theatre
- habitation – house, flat
- health – surgery, hospital
- horticulture – glasshouse
- manufacture – factory
- navigation – lighthouse
- security – prison
- social interaction – restaurant
- travel – airport, railway station
- worship – church, chapel, mosque

Such classifications carry with them certain expectations, yet in understanding a building it is important to be aware of changes and differences from what is perceived to be 'normal' (Fig. 2.7). The uses to which a building might be put, which may not necessarily be the same as that for which it was designed and built, are many and various. They may change as a building is altered or adapted; or two or more uses may be combined in one building such that a museum or art gallery combines education and entertainment, and a restaurant combines eating and social discourse.

Buildings, however, need also to be considered beyond mere type, usage and the fulfilment of basic requirements. They may demonstrate or represent creativity, offer inspiration, or arouse emotions. They may also represent the *genius loci* or spirit of the place, embracing physical, historical and aesthetic values, and giving inherent meaning and context. It is therefore necessary when attempting to appreciate a building to understand how it was made and what the designer had set out to achieve. A building might therefore be considered as one or more of the following:

Fig. 2.7   The House-in-the-Clouds, Thorpeness, Suffolk. This combination water tower and house was built in 1923 as part of the Thorpeness village development. The five-storey steel box-frame tower and oversailing two-storey superstructure, which contains the water tank, are clad with weatherboarding.

- architecture
- art
- art or architectural history
- archaeology
- landscape feature
- commodity or economic unit
- financial investment
- environmental asset (i.e. materials used and energy consumed in construction)
- cultural resource (i.e. social, political or economic)

Fig. 2.8 Anne Frank House, 263 Prinsengracht, Amsterdam. The museum, which opened on 3 May 1960, is dedicated to propagating the ideals of Anne Frank, to combating racism and anti-semitism, and to promoting a democratic, pluralistic society. Increasing visitor attendance (from 9000 in 1960 to 965 000 in 2005) led to the restoration of the house and opening of a new building to house offices, resource centre and exhibition space in 1999.

- status symbol or expression of wealth
- social statement (e.g. fashion)
- religious edifice (e.g. cathedral, synagogue)
- cultural symbol (e.g. sacred site)
- social conscience (e.g. concentration or detention camp) (Fig. 2.8)
- psychological experience
- monument
- functional machine (e.g. windmill)

Importantly, buildings may fulfil several roles – a cathedral is a building, but it is also architecture; the cathedral may contain art and it may

Fig. 2.9   The carved stone foliage and Purbeck marble shafts to the doorway into the Chapter House at Southwell Minster, Nottinghamshire (*c.* 1290) display a new artistic realism for the period in combination with conventional architectural elements. (By kind permission of the Provost and Chapter of Southwell Minster.)

also be considered, at least in part, as archaeology (Fig. 2.9). These often artificial distinctions between building, architecture, art and archaeology continue to be explored by their respective professions, but in practice often merge as architects are free to address the artistic expression of their work and archaeologists become increasingly concerned with the above-ground evidence of buildings and built forms.

Each facet of a building's existence therefore reflects and responds to different expectations and demands, and develops over time to form a complex and dynamic subject that requires detailed and often prolonged study. This is the basis for the practice of building pathology.

It is not the purpose of this book to go further in this matter, but instead six opinions on the subject are offered:

'The strength of a nation lies in the houses in which its people live.' (Abraham Lincoln, n.d.)

'I always feel sad when I look at new buildings which are constantly being built and on which millions are spent . . . Has the age of architecture passed without hope of return?' (Nicolai Gogol, *c.* 1850)

'La maison est une machine à habiter [A house is a machine for living in].'
(Le Corbusier, 1923)

'A bicycle shed is a building; Lincoln Cathedral is a piece of architecture. Nearly everything that encloses space on a scale sufficient for a human being to move in is a building; the term architecture applies only to buildings designed with a view to aesthetic appeal.' (Nikolaus Pevsner, 1943)

'Architecture is too important to be left to architects; like crime, it is a problem for everybody.' (Berthold Lubetkin, n.d.)

'On ARCHITECTURE. Buildings are not architecture. Buildings are only the means to arrive at architecture.' (Masud Taj, 1997)

The way forward

It is clear that to assess the condition of a building, and attempt to understand how it will perform in the future, consideration must be given to various levels of information. Before attempting to define what is wrong with a building, it is essential first to consider the various aspects of design and construction that will have influenced how it was built and how it has performed. Consideration of functional, performance, statutory and user requirements, and our expectations of what buildings are, will provide much that will inform such an assessment, but there are other levels of investigation that will also need to be addressed. This will include regional variations of design, material selection and utilisation, and methods of construction. In the next chapter the subject is broadened to consider the building in terms of what it is and how it performs.

References

Brand, S. (1994) *How Buildings Learn: What Happens after they're Built*. London: Viking Books.

Department of the Environment/Department of National Heritage (1994) *Planning Policy Guidance: Planning and the Historic Environment*, PPG 15, September 1994, London: HMSO.

Duffy, F. (1990) Measuring building performance. *Facilities*, **8**(5), 17–20.

Isham, G. (1995) *Rushton Triangular Lodge*. London: English Heritage.

Lovelock, J. (1995) *The Ages of Gaia*. 2nd ed. Oxford: Oxford University Press.

Osbourn, D. & Greeno, R. (1997) *Mitchell's Introduction to Building*. 2nd ed. Harlow: Addison Wesley Longman.

Further reading

Allen, W. (1995) The pathology of modern buildings. *Building Research and Information*, **23**(3), 139–46.

Brand, S. (1994) *How Buildings Learn: What Happens After They're Built*. London: Viking Books.

Day, C. (1990) *Places of the Soul: Architecture and Environmental Design as a Healing Art*. Wellingborough: Aquarian Press.

Goldstein, E.B. (1989) *Sensation and Perception*. 3rd ed. California: Wadsworth.

Gorst, T. (1995) *The Buildings Around Us*. London: E. & F.N. Spon.

Groák, S. (1992) *The Idea of Building: Thought and Action in the Design and Production of Buildings*. London: E. & F.N. Spon.

Habraken, N.J. (1998) *The Structure of the Ordinary: Form and Control in the Built Environment*. Cambridge, Massachusetts: MIT Press.

Osbourn, D. & Greeno, R. (1997) *Mitchell's Introduction to Building*. 2nd ed. Harlow: Addison Wesley Longman.

Pallasmaa, J. (1996) *The Eyes of the Skin: Architecture and the Senses*. London: Academy Editions.

Pile, S. (1996) *The Body and the City: Psychoanalysis, Space and Subjectivity*. London: Routledge.

Reid, E. (1984) *Understanding Buildings: A Multidisciplinary Approach*. Harlow: Longman Scientific & Technical.

Smith, L. (1985) *Investigating Old Buildings*. London: Batsford Academic and Educational.

Worthington, J. (1994) Design in Practice – Planning and Managing Space. In *CIOB Handbook of Facilities Management* (ed. A. Spedding), pp. 74–93. Harlow: Longman Scientific & Technical.

Chapter 3
Building Performance

Why do buildings stand up?

Although seemingly obvious, it is a fact that buildings that have stood for a period of time do not fall down unless subjected to some internal or external influence. Determining how and why a building stands, and assessing its current and future performance, is therefore a question of identifying and predicting these influences, and ensuring that changes in condition, form or usage do not bring about a critical shift in what is often a non-static equilibrium.

These influences and changes have been studied closely by Professor Jacques Heyman (1997, pp. 24–26) in relation to traditional masonry construction, with the conclusion that it is the geometry of the structure and the shape, rather than the strength, of the component materials that provide overall and long-term stability – the 'five-minute rule' confirms the success of initial design and construction, and the '500-year rule' confirms long-term fulfilment dependent upon the decay of the materials concerned (see Chapter 6).

Where a shift does take place and the equilibrium is destroyed, it is inevitable that failure will occur (Fig. 3.1). Failure, in this context, is defined as 'termination of the ability . . . to perform a required function'; a critical failure is 'that assessed as likely to result in injury to persons, significant material damage or other unacceptable conditions' (BS 3811, 1993). Investigating building or structural failures, a study sometimes referred to as 'forensic engineering', attempts to identify what happens before and during failure, and make predictions to inform the design and construction of new buildings or the nature and extent of interventions to those already standing. How a building performs is therefore of considerable importance and forms the basis for this chapter.

This performance, together with the various demands that might be implied, or imposed, by how the building is used (and abused), is perhaps the most critical consideration in assessing structural and material condition, for it is here that factors such as age, character, identity, history

Fig. 3.1 Collapse of a concrete-framed building in Mumbai as a result of ill-conceived structural alterations and lack of maintenance.

and association have to be taken into account. It is also necessary to acknowledge that materials are now being used in different ways (such as exploiting complex structural configurations) and in different situations (such as increased heating levels and thermal insulation, and decreased ventilation) than before, and that their performance has to be assessed in the light of these changes.

Buildings therefore have to be considered at various levels, which reflect the demands of their owners, occupiers and users, and of society in general. They have to be understood both as individual structures and as a collection of often diverse and disparate materials.

It is not, however, the intention of this book to consider in detail how buildings stand up as structures or perform as collections of materials. Rather, it is intended that only basic information concerning structures

and materials is given, and the reader referred to detailed texts that provide relevant coverage of the subjects. Whilst parts of this chapter may be relevant to designers responsible for the specification of materials for new buildings, it is primarily intended for those concerned with the identification of materials in existing buildings as a guide to defects (Chapter 4), assessment (Chapter 5), remediation (Chapter 6), and building management and aftercare (Chapter 7).

Building structures

The construction of buildings has typically followed a number of tried and tested patterns, based upon historical precedent and functional necessity. The primary structure of a building is thus formed either of load-bearing masonry or framed construction, with structural elements (roofs, walls and floors) forming the external enclosure, internal spaces and subdivisions, and bespoke or prefabricated components (such as ceilings, partitions, doors, windows, stairs, finishes, fixtures and fittings) added as required. Furnishings and contents may also be significant as part of an overall design concept.

The introduction of Modern Methods of Construction (MMC) offers different approaches to building in the form of off-site manufacturing (i.e. volumetric, panellised, hybrid) and non-off-site manufacturing (e.g. 'TunnelForm', 'Thin Joint Blocks'). Using prefabricated modules, panels and pods requires new constructional skills besides those of inspection and testing, and will necessitate changes in training and on-site practices.

Regardless of the method of construction chosen, or the complexity of the finished building, there are certain key determinants to the success of the construction – to disregard them typically results in increased design complexity and increased risk of failure. Essentially, the individual structural elements, the connections between them and the assembly of elements as a whole should satisfy the following criteria:

- *strength* – ability to sustain loads without undue distortion or failure
- *stability* – ability to remain balanced
- *rigidity* – ability to resist deformation under load (also referred to as stiffness)
- *equilibrium* – ability to achieve a balance of forces (static or dynamic equilibrium)
- *robustness* – ability to perform adequately for intended purposes
- *serviceability* – ability to function to satisfaction of occupants

As well as these generic requirements, the various elements, components and services that make up a building are expected to fulfil specific

needs. These are stated in the various parts of Schedule 1 to the Building Regulations 2000, and are given for reference in Appendix A.

The forms and methods of construction, in combination with the choice and application of building materials, again provide a useful insight into the processes and practices prevailing at the time of design and erection, and in subsequent alterations and changes (Fig. 3.2). Study and understanding of these will assist in determining the overall and specific performance and well-being of a building.

Fig. 3.2   The service wing of Baddesley Clinton, Warwickshire was designed and built in 1890, and blends in well with the medieval manor house despite being built of concrete blocks and sham timber framing (Norman, 1998, p. 12).

Questions to be answered in such an assessment might include:

- how was the building constructed?
- were the materials used in a conventional or an unusual manner (Fig. 3.3)?
- does the selection or manner of use of any material indicate a possible shortage in the preferred supply?

- does the form of construction follow national, regional or local traditions?
- has the construction been modified to suit specific needs?
- has a common form of construction been modified in response to a possible shortage of a principal material?
- does any material or element of construction show specific characteristics that would today be erroneously interpreted as being undesirable or defective?
- is there evidence of innovative construction practice?
- is there evidence for the use of proprietary systems of construction?

Fig. 3.3   The cupola, pediments, columns, and Corinthian decorations to the upper parts of the Court House in St Clairsville, Ohio (1886) appear to be of painted render, yet are constructed of intricately shaped and painted zinc sheet. Pressed and shaped zinc sheet is also used for internal ceilings and cornices.

Buildings might also have been designed and built to suit or exploit particular local conditions. For example:

- Temple remains from 4,000 years ago excavated at the Mesopotamian city of Mashkan-shapir, south of Baghdad in Iraq, show how large flat blocks of artificial rock were used in its construction – this stone, which is as hard and dense as basalt, was seemingly manufactured by heating local alluvial silts in large furnaces to about 1200°C, making use of technology derived from ceramic and metallurgical industries (Stone *et al.*, 1998).
- Timber-framed buildings erected from the mid-nineteenth century in the salt-working areas of Cheshire were designed so that hydraulic jacks could be placed beneath the box frames to lift them back to the required position – this allowed the buildings to withstand the effects of subsidence caused by brine pumping.
- A small number of so-called 'pot churches' built in Lancashire during the mid-nineteenth century were constructed and decorated with terracotta made from local fireclay (Hall, 1998).
- Affordable housing is being constructed in Zimbabwe in response to major housing problems through a project to improve local conditions and generate employment – low-cost alternative materials such as stabilised soil blocks (compressed blocks of soil and cement or lime) and micro-concrete roofing tiles (moulded sand and cement mortar) allow houses to be constructed with local labour and available materials (Schilderman, 1998).

Nature of building materials

From the earliest uses of mud and clay in forming daubs and sun-dried bricks to today's reliance on standardised and mechanically produced components, the construction of buildings has relied on materials to satisfy a limited number of basic requirements (Chapter 2). The choice of which material to use has, however, been dependent on several, sometimes conflicting, factors, each being set into an equation that typically attempts to balance quality, speed of building and cost.

Further factors that may influence choice to a greater or lesser extent include human responses, such as convention, personal taste and fashion, and practical issues, including the availability and quality of materials, labour, transportation, time constraints, and the imposition of restrictions, regulations and taxation.

Buildings themselves are commonly made up of various combinations of materials, brought together in ways that are derived from traditional practices or contemporary innovation (Fig. 3.4). These materials have, with the exception of a relatively small number of modern substitutes or innovations, remained constant over several hundred years, and are

typically derived from natural sources that are used either in their original recognisable state (stone, timber) or modified by one or more processes to satisfy the requirements of the client, designer, builder or regulator (brick, glass). In the case of concrete, the material can combine both structural and aesthetic qualities.

Fig. 3.4 Combinations of traditional building materials and construction practices.

Understanding building materials

Understanding the nature and limitations of traditional building materials, and thus how they were used and how they perform in the context of the built environment, requires, at the very least, an awareness of basic materials science. The chemical and physical aspects of this subject have been covered by other authors (Addleson, 1972; Torraca, 1988; Everett

& Barritt, 1994; Dean, 1996a, 1996b; Lyons, 1997), but for the purposes of this book it is sufficient to draw attention to a number of properties that characterise the durability and performance of traditional building materials.

All materials are made up of fundamental units (atoms, molecules, ions) that behave in a generally recognised manner in relation to known principles of chemistry and physics. Such principles explain the properties and behaviour of chemical elements and compounds, the forces that bind substances together, and the reactions that lead to material deterioration and eventual decay. These terms may be defined as follows:

- *element* – substance that cannot be decomposed into simpler substances
- *compound* – substance formed by the combination of elements in fixed proportions
- *atom* – smallest part of an element that can ever exist consisting of a dense nucleus of protons and neutrons surrounded by moving electrons
- *molecule* – the simplest structural unit that displays the characteristic physical and chemical properties of a compound
- *ion* – atom or group of atoms that has lost one or more electrons making it positively charged (cation) or gained one or more electrons making it negatively charged (anion)
- *chemical bond* – strong force of attraction based on transfer or sharing of electrons that holds atoms together in a molecule
- *chemical reaction* – change in which one or more chemical elements or compounds form new compounds

The durability (rate of deterioration) and performance of a material are determined by various factors that relate directly to these principles (Fig. 3.5):

- *density* – mass of substance per unit volume
- *porosity* – ratio of volume of voids to that of the overall volume of a material
- *permeability* – extent to which a material will allow a substance to pass through it
- *absorption* – penetration of one substance, such as water, into the body of another
- *adsorption* – formation of a layer of one substance on the surface of another
- *strength* (e.g. compression, tension, bending)
- *thermal properties* (e.g. conductivity, glass transition temperature)
- *acoustic properties* (e.g. transmission, sound insulation)

Fig. 3.5   Collapse of the flint facing to a clay-lump wall in south-west Norfolk. The bond between the two materials was minimal, making it susceptible to deformation and failure.

- *frost resistance*
- *soluble salt content*
- *chemical resistance*
- *fire resistance*
- *susceptibility to deformation*:
  - movement caused by applied loads
  - movements caused by changes in moisture content
  - movement caused by changes in temperature
  - stresses due to thermal and moisture changes
  - common defects due to movements and their prevention
- *susceptibility to deterioration and decay*:
  - corrosion of metals
  - sunlight
  - biological agencies (e.g. fungal, insect, vegetation)
  - water
  - salt crystallisation
  - frost action
  - chemical action
  - loss of volatiles
  - abrasion – wear or removal of the surface of a solid material due to the relative movement of another in contact with it

      ○ impact – sudden application of a load on a material
      ○ vibration – continuous and rapid movement
      ○ fire
- *natural and production defects*
- *appearance*

Sources of building materials

Most building materials are derived from natural sources governed by location (geography, geology) and local conditions (climate, exposure). These materials therefore have an affinity with their place of origin, and have thus had an important influence on the appearance and durability of buildings and structures for hundreds of years.

The visual identity of such buildings, which so readily distinguishes one part of the country from another, is also present in the ways in which such materials are put together to form buildings of practical efficiency and use. Such qualities are part of our great vernacular building tradition, and deserve to be respected in the buildings of today (Fig. 3.6).

Although most traditional building materials came from sources that were local to the building being constructed, and hence relatively easy and cheap to transport, there are occasions when materials were brought from other parts of the country or even from abroad. Such 'imports' were typically used where no suitable local materials were available, where a supply was readily obtainable, or where fashion dictated their use.

The supply and use of materials may also be understood through the influence of wider social, economic and political concerns. The shortage of good-quality materials following the Fire of London in 1666 and the requirements of the subsequent London Building Act for the laying out and construction of new buildings to avoid future conflagrations demonstrate a direct response to contemporary events. Likewise, shortages of materials and skilled labour following the Second World War and, more recently, the energy crisis of the early 1970s, led to changes in how buildings were designed, constructed and utilised.

Such reaction to national and international events will no doubt continue to influence how, where and when buildings are constructed, particularly with reference to current concerns surrounding the transfer of technologies, growing demand for housing, climate change, energy efficiency and issues of global sustainability.

Fig. 3.6   Local materials used together in a traditional manner, creating an important sense of time and place.

Concern has also been raised over current skills gaps within the construction industry, in which de-skilling, a growing emphasis on assembly rather than building, and a lack of site-based training opportunities have led to a reduction in skilled labour. This is particularly prevalent in the heritage sector, in which traditional craft skills and those associated with the supply of suitable materials are declining.

The reuse of individual building materials or components, which may be considered as a late twentieth-century response to the demands of conservation and sustainability, has been practised for many centuries. Recycled Roman brick may, for instance, be seen in many medieval buildings sited close to earlier settlements, whilst the robbing and salvage of materials, such as stone and timber, from religious houses following the Dissolution of the Monasteries in the sixteenth century helped in the building of many post-Dissolution properties.

Fig. 3.7 The medieval timber-framed Old Wellington Inn and nineteenth-century Sinclairs in Manchester following extensive bomb damage to adjacent buildings in 1996. Both buildings had been raised up by over 1.4 m during the 1970s to accommodate a reinforced-concrete deck for a shopping development, and were again moved and resited as part of a recent redevelopment scheme. (Photograph by Stephen Welsh, Buttress Fuller Alsop Williams.)

Later, in the seventeenth and eighteenth centuries, reuse was positively encouraged and even stipulated in building contracts (Beard, 1981, 26), whilst in the nineteenth and twentieth centuries whole buildings have on occasions been dismantled, moved and re-erected to satisfy individual demand or social conscience (Fig. 3.7).

The materials used in a building thus provide a useful insight into the processes and practices prevailing at the time of its design and construction, and in subsequent alterations and changes. Study and understanding of the individual and combined materials will help in forming an overall picture of how the building is performing and what action should be taken to ensure its well-being in the future. To gain such an understanding, the following questions may be asked:

- are the construction materials of local origin?
  - is there a stone quarry near the site?
  - is there evidence for pits and kilns used in the manufacture of lime, bricks, tiles, etc?
  - what are the prevalent tree species?

- what is the source of the materials (British, continental, exotic)?
- what was the connection between source and site (royal, monastic, trade)?
- how was the material brought to the site (primary goods, secondary goods, ballast)?
- is there evidence of the trade remaining (buildings, documents)?
- what processes accompanied the use of the materials?
  - were the imported materials in a finished or raw state?
  - were the materials finished on or off site (framing yards for timber frames)?
  - what were the craft skills required?

In order to develop an understanding for the use and performance of the materials utilised in the construction and finishing of buildings, it is first important to comprehend the origins and characteristics of those that are most commonly used in traditional buildings – timber, plant material, stone, ceramics, binders and concrete, metals, glass and bituminous products.

Timber

The construction and finishing of buildings has traditionally made use of many species of timber, derived from the wood of trees. The type of timber, whether native or foreign, and the ways in which it is used are both dependent on the age, location, quality and function of the building (Fig. 3.8).

Wood is defined as a hard compact mixture of cellulose, hemicellulose and lignin, forming a cellular structure that constitutes the major part of the stem or bole of a tree:

- *cellulose* – crystalline polysaccharide consisting of a long unbranched chain structure of glucose units; responsible for providing rigidity of the cell wall; represents approximately 45–60% of the dry weight of wood.
- *hemicellulose* – semi-crystalline polysaccharide consisting of a shorter chain structure; represents approximately 10–25% of the dry weight of wood.
- *lignin* – a complex amorphous organic polymer deposited within the cellulose of plant cell walls; lignification makes the walls woody and therefore rigid; represents approximately 20–35% of the dry weight of wood; greater proportion present in softwoods than hardwoods.

Fig. 3.8 Early timber-framed buildings made use of many small trees, typically oak, for both framing and finishing. In his work on the British countryside, Oliver Rackham (1986, p. 87) has calculated that some 330 oak trees were used in the construction of a fifteenth-century Suffolk farmhouse, half of the trees being less than 9 inches in diameter. These timbers, and others of greater size, might have been transported around the country, and made ready close to the location of the proposed building.

The trunk of a tree is made up of an outer bark, inner bark or bast, sapwood, heartwood and pith (Fig. 3.9). Between the inner bark and sapwood lies the cambium, which divides, usually each year in temperate climates, to form growth rings with a layer of phloem or bast (tissue that conducts food materials from regions where they are produced, such as the leaves, to the growing points) on the outside and a layer of xylem (tissue that transports water and dissolved mineral nutrients) or new wood on the inside. The heartwood, which consists of dead xylem cells that are heavily thickened with lignin, provides the main structural support for the tree. It is typically more durable and resistant to fungal attack due to the presence of oils, gums, resins and extractives (such as tannins in oak), and little free glucose or starch. The sapwood, which consists of the living xylem cells that carry sap to the leaves, is more porous than the heartwood and is typically more susceptible to fungal and insect attack.

Living trees typically hold between 100–200% moisture within the cells that make up the structure of the wood. About 25–30% of this is chemically

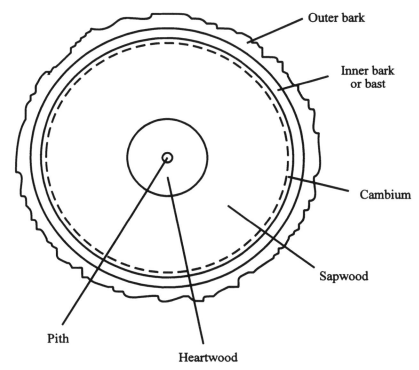

Fig. 3.9 Section through tree trunk.

bound, and the rest is present within the cells and other cavities. As timber dries, this unbound moisture is lost until only the moisture bound into the cell walls remains – this is termed the 'fibre saturation point'. Further moisture loss results in shrinkage of the cell walls and the timber, whilst the addition of moisture (due partly to the hygroscopic nature of cellulose) will cause a corresponding swelling of the timber. Timber will therefore assume an 'equilibrium moisture content' in relation to its environment.

Hardwoods and softwoods

Trees and associated cut timbers are classified as being either hardwood or softwood. Hardwoods come from broad-leaved (deciduous) trees grown in tropical or temperate climates, which bear flowers and seeds in sealed units (angiosperms), while softwoods come from coniferous (usually evergreen) trees that have naked seeds (gymnosperms) and needle-like leaves.

Although hardwood and softwood are botanical terms and do not refer to the density or hardness of a particular timber, hardwoods (such as oak,

chestnut, beech, birch and elm) are typically hard, while softwoods (such as spruce, Douglas fir, Scots pine and larch) are generally soft. By contrast, the hardwood balsa is soft, and the softwoods pitch pine and yew are hard. Each timber should therefore be identified by family, genus and species for the purposes of specification or supply. Hardwoods and softwoods commonly used in the construction and fitting-out of buildings are given in Table 3.1.

Table 3.1 Examples of commonly used hardwoods and softwoods.

Common name	Latin name	Country of origin
Hardwoods		
Ash	*Fraxinus excelsior*	Europe
Beech	*Fagus* spp.	Europe, Japan
Birch	*Betula* spp.	Europe
Chestnut (Sweet)	*Castanea sativa*	Europe
Elm	*Ulmus* spp.	Europe, USA, Japan
Mahogany (African)	*Khaya* spp.	West Africa
Mahogany (American)	*Swietenia macrophylla*	Central and South America
Oak	*Quercus* spp., *Q. robu* and *Q. sessliflora*	Europe, America, Japan
Softwoods		
Douglas fir	*Pseudotsuga taxifolia*	Canada, USA
Parana pine	*Araucaria angustifolia*	Brazil
Pitch pine	*Pinus palustris, P. caribaea* and spp.	USA, Central America
Scots pine (Redwood)	*Pinus sylvestris*	Europe
Western white pine	*Pinus monticola*	Canada, USA
Whitewood (European spruce)	*Picea abies*	Europe
Yellow pine	*Pinus strobus*	East Canada

Differences in cellular structure between hardwoods and softwoods will influence:

- strength, weight, buoyancy, density
- hardness
- fire resistance
- planed appearance and polish (reflectivity)
- ability to be impregnated with preservatives, paints or adhesives
- susceptibility to moisture ingress
- susceptibility to beetle and fungal attack
- identification and classification of timbers

Seasoning and conversion

Before timber can generally be used it must first be seasoned to reduce its equilibrium moisture content to that of its intended environment (Equation 3.1). Timber used for external applications may thus have a moisture content of 16–18%, whilst that used internally may need to be kiln seasoned to 8–14% moisture content in order to avoid shrinkage resulting from higher ambient temperatures. Timber with a moisture content of above 20% is susceptible to fungal attack.

$$\text{Moisture content} = \frac{(\text{wet weight of sample}) - (\text{dry weight of sample})}{(\text{dry weight of sample})} \times 100\% \qquad (3.1)$$

Unseasoned or 'green' timber has, however, traditionally been used for the fabrication of timber frames as it is easier to cut and work, and is one of the reasons why timber-framed buildings may be seen to have distorted without apparent loss of stability. Flexible joints held together with wooden pegs allow the various timber members to move as they season *in situ* and respond to later fluctuations in temperature and humidity (Fig. 3.10). Smaller timber sections used for making the woven wattle framework of infill panels or the fixings for thatch roof coverings are also

Fig. 3.10 Fifteenth-century timber-framed barn at Newton Flotman, Norfolk showing fine double queen-post roof. (Photograph by Stephen Heywood.)

used unseasoned to provide initial flexibility, and are often harvested from coppiced or pollarded trees.

Timber used for carpentry or joinery is converted according to particular structural, dimensional or visual requirements. Trees with few knots and a straight grain might originally have been split with wedges, whilst others would have been rough sawn and shaped with an adze (a form of axe). Sawn timber has traditionally been prepared either by *tangential* or *radial* sawing, whilst recent developments in *star-cutting* offer increased efficiency and less wastage (Tickell, 1998, p. 10) (Fig. 3.11).

Timber is dimensionally unstable, and shrinkage is not uniform along the three principal axes (anisotropic). Movement is therefore almost twice as great for timber cut tangentially than for radially cut timber. This can cause distortion (twisting, bowing, cupping) of the timber in use and may be seen with roofing shingles and internal fittings.

Modern timber products

Modern timber products (laminated timber; engineering timber; plywood; hard-, medium- and softboard; blockboard or laminboard; chipboard or particleboard; orientated-strand board; cement-bonded particleboard; fibreboard; woodwool slabs) provide alternative and specific materials for construction and finishing. Innovative use is also being made of small-diameter forest thinnings for the creation of bent lattice structures in conjunction with turf roofs.

Plant material

The stems of plants such as reed, rushes and cereal crops, together with other fibrous plant growth (bracken, broom, flax, gorse, heather, marram grass, sedge and sods), have been used throughout history to provide a lightweight and durable thatch to roofs (Fig. 3.12). Such coverings rely typically on a steep roof pitch to quickly shed rainwater and secure fixings to avoid damage by strong winds (Fig. 3.13), and, more recently, chemical treatments to provide fire resistance. Reed and straw may also be used to form coverings to wall frames, and as the support for plaster floors.

Smaller plants and vegetable material (such as grasses, mosses and the roots of larger plants) have also been used to provide reinforcement in clay and earth construction, and to act as insulation beneath roof coverings. Straw, jute and hemp have similarly been used to provide reinforcement in lime and gypsum plasters.

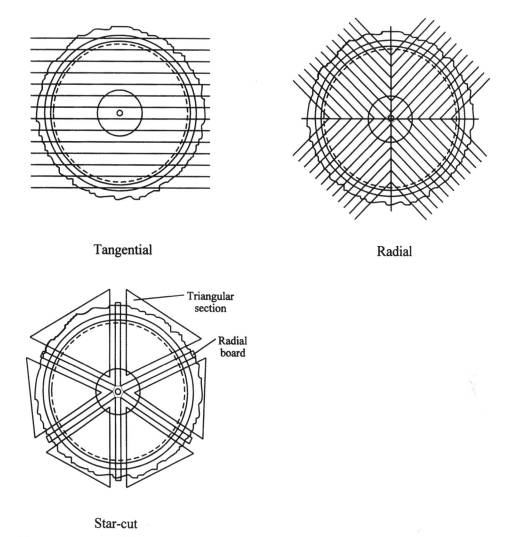

Tangential

Radial

Star-cut

Fig. 3.11 *Tangential, flat or plain sawn* timber is sawn tangentially to the annual growth rings (i.e. the growth rings meet the face in any part at any angle of less than 45°); it is quicker to cut and dry and there is less waste. *Radial or quarter sawn* timber is sawn perpendicular to the growth rings (i.e. the growth rings meet the face at an angle of not less than 45°); it dries more slowly and is dimensionally more stable. *Star-cut* timber is sawn perpendicular to the growth rings, with the triangular sections being glued to make dimensionally stable panels.

Modern plant material products

Growing interest in 'green' and sustainable buildings has led to a corresponding supply of natural building materials that can offer a realistic alternative to the energy-intensive production of artificial mass-produced products. This includes the use of thatch on new buildings (Boniface, 1998).

Fig. 3.12 Cutting water reed at Hickling in Norfolk.

Fig. 3.13 Long-straw thatching in progress using specifically grown materials to avoid deterioration caused by chemical fertiliser residues.

Demand is at present being met by a small, but growing, number of suppliers who are offering both alternatives to current products and new forms of construction. Such uses for plant material include the production of compressed straw boards and blocks, and straw-bale construction (Fig. 3.14).

Fig. 3.14   The Spiral House in County Mayo, Ireland is Europe's first two-storey, load-bearing straw-bale house and was constructed by Amazon Nails and volunteers in 2000–01. (Photograph by Amazon Nails.)

Stone

In order to understand how building stones react and respond to mechanisms of deterioration and decay, it is first necessary to appreciate what the stones are and how they were formed. It is therefore essential to be aware of the geology (study of the crust and strata of the Earth) and petrology (study of the origin, structure and composition of rocks) of the parent rocks.

Geological classification

Rocks may be classified by their geological terms, based upon the era and period in which they were formed (Table 3.2). The British Geological Survey 'Rock Classification Scheme' provides an integrated approach

Table 3.2 Geological classification (ages in millions of years).

Igneous rocks (mode of origin)
- Intrusive – mainly granite, granodiorite, gabbro and dolerite
- Volcanic – mainly basalt, rhyolite, andesite and tuffs

Metamorphic rocks
- Lower Palaeozoic and Proterozoic (500–1000) – mainly schists and gneisses
- Early Precambrian (Lewisian) (1500–3000) – mainly gneisses

Sedimentary rocks (age of deposition)
- *Cenozoic era (recent life)*
  - Tertiary and marine Pleistocene (up to 65) – mainly clays and sands
- *Mesozoic era (middle life)*
  - Cretaceous (65–140) – mainly chalk, clays and sands
  - Jurassic (140–195) – mainly limestones and clays
  - Triassic (195–230) – marls, sandstones and conglomerates (New Red sandstone)
- *Palaeozoic era (ancient life)*
  - Permian (230–280) – mainly magnesian limestones, marls and sandstones (New Red Sandstone)
  - Carboniferous (280–345) – limestones, sandstones, shales and coal seams
  - Devonian (345–395) – sandstones, shales, conglomerates (Old Red Sandstone), slates and limestones
  - Silurian (395–445) – shales, mudstones, greywackes, some limestones
  - Ordovician (445–510) – mainly shales and mudstones, limestone in Scotland
  - Cambrian (510–570) – mainly shales, slate and sandstones; limestone in Scotland
- *Upper Proterozoic*
  - Late Precambrian (600–570) – mainly sandstones, conglomerates and siltstones

---

to classification based on description rather than interpreted attributes, and is presented in four volumes covering igneous rocks, metamorphic rocks, sediments and sedimentary rocks, and artificial man-made ground and natural superficial deposits.

Geological maps

Geological maps produced by the British Geological Survey are of two main types:

- bedrock geology (formerly known as 'solid') maps that represent rocks deposited or created before the Quaternary Ice Age (prior to 1,770,000 years ago)
- superficial deposit (formerly known as 'drift') maps that show deposits laid down during the Ice Age and the present post-glacial epoch

The 'bedrock' maps represent Pre-Quaternary rocks as if the Quaternary sediments have been removed, and the 'superficial' maps represent deposits as they occur at the land surface.

Computer-based map and data services are also available, which provide information on land usage and ground conditions. Landmark Information Group Limited provides an account of land use for the period *c*.1840 to *c*.1995 using historical map and land-use data sets with the integration of almost one million Ordnance Survey sheets. The British Geological Survey can supply *GeoReports* for most of Scotland, Wales and England, either in the form of a standard geological assessment or a natural ground stability report that make use of records and maps held in the National Geoscience Data Centre. The British Isles GPS Archive Facility (BIGF) continuously records global positioning system (GPS) data at over 100 permanent stations throughout Britain, which has practical application in various environmental disciplines (e.g. archaeology, coastal engineering, flood risk and assessment, national infrastructure).

Petrographic classification

The parent rocks that provide stone for building may be classified as igneous, sedimentary or metamorphic. These terms describe the origins of the rocks and are therefore descriptive of the ways in which they were formed.

Igneous rocks (including granite and basalt) are formed by the solidification of molten rock material, whether within the earth's crust (plutonic) or at the surface (volcanic). Most are composed of a few mineral groups (quartz, feldspars and feldspathoids, pyroxenses, amphiboles, micas and olivines), and can be considered as being either coarse (>1–2 mm, such as granite), medium (0.1 to 1–2 mm, such as micro-granite) or fine grained (<0.1 mm, such as basalt) (Fig. 3.15).

Sedimentary rocks (including limestones and sandstones) are formed by the accumulation of rock waste at the Earth's surface, in contrast to igneous and metamorphic rocks produced by internal processes within the Earth. Such rocks are considered as coarse- (>2 mm, such as gravel), medium- (2 to $\frac{1}{16}$ mm, such as sand) or fine-grained ($\frac{1}{16}$ to $\frac{1}{256}$ mm, such as silt, and $<\frac{1}{256}$ mm, such as clay) (Fig. 3.16), and may be classified according to their origin – sediments transported as solid particles by water, wind or ice (mechanical), sediments formed by precipitation from solution of dissolved salts and sometimes by chemical replacement of one mineral by another (chemical), or sediments formed by the accumulation of organic material, whether animal of plant (organic) (Table 3.3).

Metamorphic rocks (including slate and marble) are formed through the alteration of igneous and sedimentary rocks by the action of heat and

Fig. 3.15   The igneous and metamorphic rocks of the Charnwood Forest area of Leicestershire have provided granite and slate for a number of local buildings.

Fig. 3.16   Clunch, a compacted type of chalk from the Cretaceous formation in East Anglia, being quarried at Barrington, Cambridgeshire.

Table 3.3 Classification of sedimentary rocks according to their origin.

Mechanical origin
- Coarse – conglomerate
- Medium – sandstone
- Fine – siltstone, mudstone, shale

Chemical origin
- Calcareous (containing calcium carbonate) – calcareous mudstone, oolitic limestone (in part), travertine
- Dolomitic (containing magnesium and calcium carbonates) – dolomite
- Siliceous (containing silica) – flint, chert
- Ferruginous (containing iron oxide) – ironstone, carstone
- Argillaceous (containing clay) – septaria
- Saline – rock salt, gypsum rock
- Phosphatic – phosphate rock (in part)

Organic origin
- Calcareous – biochemical limestone, oolitic limestone (in part)
- Carbonaceous – coal
- Phosphatic – phosphate rock (in part)

pressure, the most common being slate, phyllite, schist, gneiss and hornfels (see Fig. 3.15). Parent rocks such as quartz and sandstone will form quartz schist and quartzite, greywacke will form schist, pure limestone will form marble, impure limestone will form calcareous schist and calcareous hornfels, shale and mudstone will form slate and hornfels, and diabase/basalt will form greenschist and basic hornfels.

Identification and description of rocks and stones

Identification and description of rocks and building stones requires consideration of grain size, texture, structure, mineralogy and colour (lithology). In the case of sedimentary rocks, attention should in particular be given to the shape, texture and range of the grains; specific features such as the inclusion of shells, fossils and oolites (small round or ovoid bodies with a concentric structure of layers of calcite around a nucleus); and the pattern of the bedding planes.

An example of a geological description for Weldon stone based on such observations would be:

'Porous shelly oolitic limestone; cross-lamination often visible, defined by oyster-rich layers. Workable as a freestone; available in large blocks.' (Hudson & Sutherland, 1990, p. 23)

Examples of common building stones are given in Table 3.4.

Table 3.4 Common building stones.

Name of stone	Location of quarry	Texture and colour
Limestones		
Ancaster	Grantham, Lincolnshire	Fine-grained with varying shell content, cream to buff
Anston	Kiveton, Sheffield	Fine-grained, yellow to cream
Barnack	Stamford, Lincolnshire	Coarse textured, shelly, cream-buff
Bath stones	Avon	Even-grained, pale brown to light cream
Casterton	Stamford, Lincolnshire	Coarse-grained, beige
Clipsham	Oakham, Rutland	Medium-grained with shelly fragments, buff to cream, some blue heart
Doulting	Somerset	Coarse-grained, pale to dark brown
Ham Hill	Hamdon Hill, Somerset	Coarse-grained, yellow to rich brown
Hopton Wood	Derbyshire	Fine-grained, cream
Ketton	Stamford, Lincolnshire	Medium-grained, oolitic, yellow to buff, some distinct pink
Portland stones	Dorset	Even-grained with shell matter and open texture, pale brown to buff
Purbeck-Portland	Isle of Purbeck, Dorset	Shelly, blue/grey to buff
Sandstones		
Birchover	Stanton Moor, Derbyshire	Medium- to coarse-grained gritstone, pink to buff
Darley Dale	Matlock, Derbyshire	Fine-grained, compact pale brown to white
Elland Edge	Brighouse, West Yorkshire	Fine-grained, fissile, brown to grey
Hollington	Uttoxeter, Staffordshire	Fine- to medium-grained, white, red and mottled
Mansfield	Nottinghamshire	Fine-grained, white, yellow or red
St Bees	Cumbria	Fine-grained, red to brown
Stone slates		
Limestone		
Collyweston	Northamptonshire, Rutland, Lincolnshire, Cambridgeshire	Smooth textured, bluish weathering to buff
Purbeck	Isle of Purbeck, Dorset	Grey, blue-grey and browns
Stonesfield	Oxfordshire	Medium textured, pale creamy yellow weathering to brown
Sandstone		
Elland	Brighouse, West Yorkshire	Gritty texture, dark grey to brown
Hoar Edge	Cardington, Shropshire	Coarse-grained, shelly, buff to brown
Horsham	Surrey, Sussex, Kent	Often ripple-marked, buff to brown

Contd.

Table 3.4 Contd.

Name of stone	Location of quarry	Texture and colour
Metamorphic		
Ballachulish	Argyll	Coarse texture, blue-grey to black
Delabole	Camelford, Cornwall	Coarse texture, grey weathering to grey-green
Swithland	Leicestershire	Fine texture, blue-grey to green
Welsh	Bethesda, Llanberis, Ffestiniog	Fine texture, various shades of blue, blue-purple, grey and green
Westmorland	Cumbria	Fine texture, various shades of green and blue-grey

Modern stone products

Artificial or reconstituted stone provides an economical alternative to the use of natural stone both for new construction and for the repair of existing buildings. Such stone is typically cast as a mixture of stone dust, natural aggregates and cement, and used either as a solid block or as a facing to a backing material such as concrete. Reconstituted slate and marble are based on granules or chippings of the original material, together with fillers and resins.

Ceramics

The term 'ceramic' (from the Greek *keramos*, meaning 'burnt stuff') describes both the use of raw plastic clay and the resulting manufactured goods, and can refer to either clay products (such as bricks, tiles and pipes) or pottery products (such as terracotta, faience, majolica, earthenware, vitreous china, stoneware, porcelain and bone china) (Fig. 3.17). Each is based on the use of certain clays (primary or secondary) and other materials, methods of preparation, firing temperatures, and specific finishes. The term may, in certain circumstances, also refer to unfired mortars, plasters, daubs, mud blocks and concrete (Middleton, 1991, pp. 17–21).

The clays used in the manufacture of ceramics, many of which give characteristic colours to regional bricks and tiles, are composed of silica (40–65%) and aluminium oxide (alumina) (10–25%), with various impurities including iron compounds, magnesia, potash, soda, lime and sulphur. These clays are typically extracted from quarries or pits, prepared, blended (if necessary), moulded (or mechanically extruded or pressed) to shape, dried and fired to create the basic unit.

Fig. 3.17 The frontage to the Wedgwood Memorial Institute in Burslem, Staffordshire, was designed following a competition based on the use of decorative ceramics. The chosen scheme was by Robert Edgar and John Lockwood Kipling, which proposed a polychromatic design of decorated brickwork, tiles, terracotta mouldings and panels, mosaic, and Della Robbia ware. Terracotta panels show the months of the year, with mosaic signs of the zodiac above, and bas-relief panels depict the processes of the pottery industry. The building was completed in 1873, when a terracotta statue of Josiah Wedgwood was fixed above the entrance.

Bricks and tiles

Brick-making has an ancient tradition dating back about 5000 years in the Near East – these bricks, whether of mud or clay, were the earliest building material to be manufactured. Clay bricks and tiles were first introduced to Britain during the Roman occupation, and re-established with the help of continental craftsmen in the late twelfth century. Both became principal materials in urban building due to the growing demand for fire-resistant construction (such as required by the Building Act of 1667 following the Great Fire of London in the previous year), and have remained in constant use to the present time (Fig. 3.18).

The size and quality of bricks have varied throughout history, as a result of developing production methods and the demands of a maturing building industry. Roman bricks were, for instance, typically long and thin (such as $12 \times 6 \times 1^{1}/_{4}$ in), whilst those of the fifteenth century averaged $9 \times 4^{1}/_{2} \times 2$ in (Clifton-Taylor, 1987, 213). Various statutes have imposed

Fig. 3.18 Brick chimney stacks, plain roof tiles and ornamental tile hanging combine with timber framing in Marlborough, Wiltshire to illustrate the influence of changing tastes and fashions.

restrictions on size, and the Brickmaker's Charter of 1571 established what is called the Statute Brick of $9 \times 4^{1}/_{2} \times 2^{1}/_{4}$ in. Mechanisation in the nineteenth century and metrication in the 1970s has since forced standardisation and led to a loss of local distinctiveness.

Improved methods of production (such as kiln design, and higher and controllable firing temperatures) and choice of clays have resulted in various brick types, with classification based on the following criteria:

- *place of origin* – Fletton (Cambridgeshire), London stock, Staffordshire blue engineering brick
- *composition of clay* – Keuper marl, Etruria marl, Oxford clay, London clay, Coal Measure shale

- *type of brick* – solid, perforated, frogged, hollow, glazed, specials
- *specific application* – facings, commons, rubbers, engineering, refractory
- *appearance* – colour, surface texture
- *durability* – frost resistance, soluble salt content
- *physical properties* – strength

Bricks may be used to form solid or cavity walls, or a facing to a structural backing wall, and include damp-proof courses (introduced by the Public Health Act of 1875), air bricks, lintels, cavity ties and trays as part of the construction. Brick bonding patterns (such as English bond and Flemish bond), used to provide strength and improve appearance, together with surface patterning (diaperwork) and mixed colours (polychromy), are often indicative of the age and original importance of a wall (Fig. 3.19).

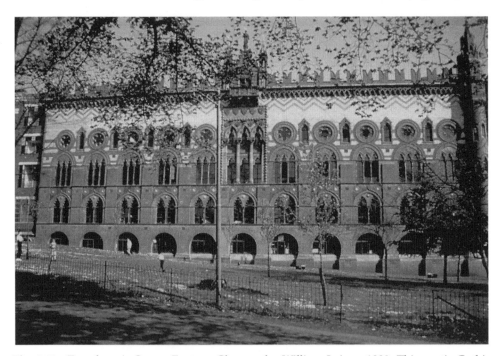

Fig. 3.19 Templeton's Carpet Factory, Glasgow by William Leiper, 1889. This exotic Gothic building is one of the many British imitations of the Doge's Palace in Venice. (Photograph by Peter Swallow.)

In addition to conventional brickwork, mathematical tiles were used as a wall cladding to houses in the south-eastern counties of England during the eighteenth and nineteenth centuries. The use of such tiles, in imitation of brickwork, may have been stimulated partly by the imposition of taxes on brick (introduced in 1784, increased in 1794 and 1805, and repealed in 1850), from which they were exempt.

Calcium silicate (or sand-lime) bricks, manufactured from a mix of silica (sand), hydrated lime, crushed flint, coloured pigments and water, have been available in Britain since 1905 and provide an alternative to clay bricks for internal or external usage.

Terracotta and other ceramic materials

Moulded blocks of unglazed terracotta were first imported into Britain from Italy in the early sixteenth century, and manufactured at local centres from the eighteenth century. This material was used extensively in both structural (either as hollow or concrete-filled blocks) and decorative roles in towns and cities during the nineteenth century because of its durability and suitability for repetitive detailing (Fig. 3.20).

Other ceramic materials include:

- *Coade stone* – hard, durable material resembling stone used for decorative features from late eighteenth century
- *earthenware* – glazed for use in the manufacture of drainage goods
- *faience* – glazed terracotta used in structural units or as decorative slabs from late nineteenth century

Fig. 3.20 Hand finishing a decorative terracotta block at Ibstock Hathernware Limited, Leicestershire.

- *fireclay* – fire-resistant material used for grate blocks and flue liners
- *glazed tiles* – glazed earthenware used for decoration and protection
- *stoneware* – unglazed for use in the manufacture of drainage goods and cladding panels
- *vitreous china* – impermeable material used in the manufacture of sanitary goods

Clay and earth construction

Clay and earth have been used in the construction of buildings for many centuries. Regional variations are particularly important, and have received growing interest in recent years.

The main techniques used in the formation of load-bearing walls are:

- *cob* – made from clay, straw and aggregate, and built up without shuttering in lifts usually off a masonry plinth; found principally in the south-west of Britain, but also in the East Midlands (mud-walling) and Cumbria (clay daubing)
- *clay lump* – unfired moulded block of chalk, clay and straw found particularly in East Anglia from the late eighteenth century to early twentieth century; the term 'adobe' is used in other countries (such as Spain, Mexico and North America) to describe a similar form of construction
- *clom* – a local technique from Wales based on a mix of clay, aggregates and chopped wheat straw (or chaff, rushes, bracken, moss or animal hair) laid in layers on to a masonry plinth and trimmed to a smooth battered finish
- *pisé de terre* – introduced in the late eighteenth century from Continental Europe, and formed by compacting layers of suitable dry earth between shutter boards; also known as rammed earth
- *pugged chalk* or *chalk-mud lump* – a mix of soft chalk and clay used either with shuttering or as pre-cast blocks in areas of natural chalk deposits
- *shuttered clay* – similar to pisé, but using chalk, clay and straw, and found particularly in East Anglia
- *witchert* – a mix of earth and straw built up from a plinth with no shuttering; found principally in Buckinghamshire; also spelt wychert and witchit

Clay and earth may also be used to form panels between the members of external or internal timber wall frames. Such techniques, which have distinct regional variations, include:

- *clam-staff and daub* – clay-based infill built up around thin studs in buildings on the Lancashire plain

- *mud and stud* – lightweight timber frame completely covered with a thick daub, used particularly during the late seventeenth and early eighteenth centuries in the Lincolnshire Wolds
- *wattle and daub* – common form of infill with clay- or lime-based daub applied on to pliable sticks or reeds woven around vertical staves (Fig. 3.21)

Fig. 3.21 Construction of new wattle-and-daub panels.

Such forms of construction typically rely on the protection of a masonry plinth, overhanging eaves, porous wall coverings (plaster, render) and appropriate surface finishes (limewash). Often such renders and plasters are themselves based on clay rather than lime, and may be seen in East Anglia, Ireland and Scotland.

Clays and earths, often with sands, gravel and other inert materials, have also been used to form cheap and functional floors throughout history. These are usually found in rooms and spaces of older buildings that do not have heavy traffic and are not prone to wetting, and are typically beaten or rammed to a smooth and compact finish.

Modern ceramic products

Modern ceramic products and practices (reinforced brickwork, pre-assembled brickwork, concrete bricks and blocks, fair-faced blockwork) provide alternative and specific materials for construction and finishing. Certain historic materials, such as clay and earth, are also being used in a modern context to provide economic and environmentally sustainable forms of construction – stabilised soil blocks, formed by compressing a mix of soil and cement or lime into moulds, are being used in various countries around the world to provide a low-cost material for simple forms of construction.

Binders and concrete

The term 'binder' describes those materials used to form the cementing matrix in mortars, renders and plasters, and which typically are derived from lime, gypsum or artificial cements. Such binders are manually or mechanically mixed with sands and aggregates, and sometimes with other materials (such as hair, straw, blood, sugar, milk or egg-whites), to provide a workable product that can be used between masonry units, as an applied finish to walls and ceilings, for the casting and running of decoration ornamentation, and to form plaster floors. Other binders, such as clay and bitumen, may also be used for specific applications.

Concrete, by comparison, refers to a mix of binder (whether lime or cement) with sands and other aggregates to form an amorphous material capable of use, whether with reinforcement or not, in a variety of structural and non-structural applications.

The proportion of binder to aggregates and the amount of water used in a mix for mortars, surface finishes or concrete are derived from practice or design to ensure that all the particles are coated and the voids between the particles filled. This ensures that the final product is both economical and durable.

Binders

Lime used in the construction and finishing of buildings is derived from the burning of limestone, chalk or other sources of calcium carbonate (coral, sea shells). The resulting calcium oxide or quick lime is slaked with water to form calcium hydroxide, which in use takes up carbon dioxide from the atmosphere to reform as calcium carbonate. These are sometimes referred to as 'air limes'. This cycle of lime production and use is explained in Equation 3.2.

(1) *Burning*:
calcium carbonate + heat (950°C) → calcium oxide + carbon dioxide

$CaCO_3$ + heat   → CaO + CO_2

(2) *Slaking*:
calcium oxide + water → calcium hydroxide

CaO + H_2O → $Ca(OH)_2$

(3) *Carbonation*:
calcium hydroxide + carbon dioxide → calcium carbonate + water

$Ca(OH)_2$ + CO_2 → $CaCO_3$ + H_2O (3.2)

> Where predominantly argillaceous or siliceous limestones are burnt and reduced to powder, the resulting natural hydraulic lime (NHL) has the ability to set and harden under water. A hydraulic lime (HL) is produced by mixing materials consisting of calcium hydroxide, calcium silicates and calcium aluminates, and can also set and harden under water. Atmospheric carbon dioxide contributes to the hardening process in both cases. Such limes have been widely used throughout history for general building and engineering works. The classification of hydraulic action relating to clay content – feebly hydraulic (0–8%), moderately hydraulic (8–18%) and eminently hydraulic (18–25%) – has been replaced by a system based on compressive strength after 28 days (e.g. HL 3.5 or NHL 3.5 will have a compressive strength of >3.5 to <10 MPa after 28 days).
>
> In regions where such limes were not available or where hydraulicity needed to be increased, it has been traditional practice to add pozzolanic or hydraulic materials to form artificial hydraulic limes (termed 'NHL with additional material'). Pozzolana, named after the volcanic ash found at Pozzuoli near Mount Vesuvius in Italy, includes both natural materials such as 'trass' from the Upper Rhine and 'santourin' from Greece, and artificial fine-powder materials such as low-fired brick dust, pulverised fuel ash and high-temperature insulation powder. These contain reactive silica (SiO_2) and alumina (Al_2O_3) that react with the slaked lime ($Ca(OH)_2$) to produce calcium silicate hydrate. This forms a network of fibrous crystals or gel, which causes the binder to harden and develop a hydraulic set.
>
> Artificial binders based on hydraulic lime and patented oil mastics or cements became popular during the eighteenth and nineteenth centuries, typically for use as renders and stuccoes and often in imitation of more expensive stone. Parker's Roman Cement (based on the burning of argillaceous limestone and patented 1796) was commonly used until the widespread acceptance of Portland cement in the early nineteenth century, whilst Dehl's mastic (patented 1815) and Hamelin's mastic (patented 1817) were similarly applied to form attractive and durable wall finishes.

Portland cement, named on account of its supposed resemblance in colour to Portland stone, was patented by Joseph Aspdin in 1824, and has become almost universal in its application as a binder and for the formation of concrete. The chemistry of cement is complex, but in principle relies on the reaction of four main compounds derived from the clay or shale and limestone or chalk (tricalcium silicate, dicalcium silicate, tricalcium aluminate and tetracalcium alumino ferrite) with water. This hydration process leads to the formation of an alkali paste that slowly stiffens and hardens with the formation of crystalline products (ettringite, calcium silicate hydrate, tricalcium aluminate, calcium hydroxide) between the cement grains (Equation 3.3 using cement notation).

tricalcium silicate + water → calcium silicate hydrate + calcium hydroxide

$2C_3S$ $\qquad + 6H \quad → C_3S_2H \qquad\qquad + 3CH$

dicalcium silicate + water → calcium silicate hydrate + calcium hydroxide

$2C_2S$ $\qquad + 4H \quad → C_3S_2H_3 \qquad\qquad + CH$ $\qquad\qquad$ (3.3)

Variations of Portland cement based on differing proportions of the above four compounds include ordinary, rapid-hardening, ultra-high early-strength, sulphate-resisting and low-heat Portland cements. Other types of cement include white cement, masonry cement, Portland blast-furnace cement, super-sulphated cement and calcium aluminate (high alumina) cement.

Gypsum (hydrated calcium sulphate), often termed 'plaster of Paris' on account of the gypsum quarries at Montmartre in Paris, has also been used historically as a binder for internal plasters, and particularly for moulded decorations owing to its quicker setting time over lime. As gypsum can be fired at a lower temperature than lime (Equation 3.4), it is often preferred in countries where fuel is scarce.

hydrated gypsum + heat (130°C) → hemi-hydrate + heat (160°C) → anhydrous gypsum

$CaSO_4.2H_2O$ $\quad + $ heat $\qquad → CaSO_4.\frac{1}{2} H_2O + $ heat $\qquad → CaSO_4$ $\qquad$ (3.4)

Artificial products, such as Martin's Cement (patented 1834), Keen(e)'s Cement (patented 1838) and Parian cement (patented 1846), were typically based on gypsum and used to form high-strength plasters for finishing and mouldings. Today, gypsum forms the basis for most internal plasters and plasterboard products.

Clay binders have also been used for renders and plasters, particularly to form a finish on clay and earth construction, and also to form masonry mortars in certain parts of Britain.

Concrete

Early concrete used in Roman construction was formed as an aggregate of pozzolana, sand, gravel and rubble material, and laid in a matrix of lime mortar. This material was typically used in foundations and as a filler in masonry construction, and also for the formation of arches, domes and vaults. The development and use of modern concrete construction began, however, with the work of Joseph Aspdin in the late eighteenth century and the patenting of Portland cement in 1824 (see above).

Concrete has good compressive strength, but is comparatively weak in tension and requires reinforcement to allow it to withstand tensile forces. Basic reinforcement was added from the early nineteenth century – patents included the inclusion of wrought-iron bars (1818), floor construction using iron girders and concrete infill (1844), the use of iron mesh (1848), fireproof construction using concrete reinforced with wire ropes (1854) and a basic form of reinforced concrete (1855) – but it was not until the early twentieth century that 'ferro-concrete', developed by François Hennebique and combining the qualities of concrete and steel reinforcement, was recognised and taken up by architects and engineers (Fig. 3.22).

Fig. 3.22   The graceful four-arch US Route 40 concrete viaduct of the mid-1930s in Belmont County, Ohio overshadowing the Blaine Hill 'S' Bridge built as part of the national road project in 1828.

The earliest recorded concrete house was built in Swanscombe (Kent) in 1835, whilst the first large reinforced-concrete building was Weaver & Company's mill in Swansea (1897–98). Further examples of early innovation

using concrete include the high-rise Royal Liver Building in Liverpool (1908) and the shell roofs of the Gare de Bercy in Paris (1910) and Jahrhzunderthalle (Centenary Hall) in Wroclaw, Poland (1911–13). Today pre-stressed concrete (pre-tensioned or post-tensioned), as developed by Eugène Freyssinet in the 1920s, is commonly used for both building and engineering applications.

Pre-cast concrete construction, used for structural members, cladding panels and individual building units, was similarly adopted in the early years of the twentieth century, and has become widespread both for building construction and engineering applications.

Modern binders and concrete

The basic materials of lime and artificial (typically Portland) cement continue as the principal binders in the construction of buildings, whilst the use of hydraulic and non-hydraulic limes for the repair and maintenance of traditional and historic buildings receives renewed support as more becomes known about the importance of 'soft' construction and water-vapour permeability. Different forms of pozzolana, such as brick dust and micro-silica cements, are also being investigated for their influence on the performance of masonry mortars. Modern gypsum plasters and plasterboard products have achieved universal application, with special gypsum-based plasters available to suit particular applications (renovating, projection, acoustic, X-ray, textured and fibrous plasters).

Cementitious and non-cementitious grouts, both for building and engineering works, have also been developed for specific applications. These include general-purpose grouts, low-strength grouts (such as those based on a blend of lime, pulverised fuel ash and clay for filling voids in rubble-core walls), polymer-modified grouts for application in wet conditions or where the initial set needs to be controlled, flow-modified grouts and accelerated grouts.

Structural concrete, whether pre-cast or cast *in situ*, has become universally known and used for buildings and engineering structures, and as a popular material for designers wishing to explore its unique capacities through complex three-dimensional forms (Fig. 3.23). Modern pre-cast concrete blocks and bricks, using a variety of aggregates and manufacturing processes, have also achieved widespread application both for loadbearing construction and as non-loadbearing internal partitions.

Metals

Metals used in the construction of buildings may be divided between those based on iron (ferrous) and those of an alternative (non-ferrous)

Fig. 3.23   High Cross House, Dartington Estate by William Lescaze (1932). (Photograph by Susan Macdonald.)

composition. Ferrous metals include cast and wrought iron and various forms of steel; non-ferrous metals include aluminium, copper, lead and zinc. Internal decorative finishes and contents may also make use of a wider range of metals, including brass, bronze, gold, silver, pewter and tin.

Ferrous metals

Ferrous metals are composed principally of iron with varying amounts of carbon and other elements. The resulting materials have differing characteristics and uses, although steel has become the principal structural metal for construction during the twentieth century.

Wrought iron was first produced in Britain in *c.* 450 BC using a technique of single-stage direct reduction, where iron oxide was heated in contact with carbon (Equation 3.5) using 'bloomeries' (small furnaces that produced small lumps or 'blooms' of iron). The availability of wrought iron increased with the introduction of two-stage indirect reduction in *c.* 1400, where iron ore was smelted to produce 'pig iron', which was then converted to wrought iron by the removal of carbon.

Iron ore + carbon → iron + carbon monoxide

$$Fe_2O_3 + 3C \rightarrow 2Fe + 3CO \tag{3.5}$$

Wrought iron is a relatively pure material (Fig. 3.24), with typically less than 0.03% carbon in its composition, and has a fibrous structure making it strong in tension. It can be worked to shape by hammering, squeezing or rolling, this being done whilst the metal is hot, although cold working is also possible. The production of wrought iron has ceased, and fabrication is today limited to the use of recycled material.

Cast iron was introduced into Britain during the sixteenth century and, with better techniques of smelting, became widely used in heavy engineering during the nineteenth century. It is essentially an alloy of iron and carbon, with up to 5% carbon content. If the carbon is present as graphite, the metal is termed 'grey' cast iron; if it is present as carbide, it is termed 'white' cast iron. The resulting metal is crystalline in structure and brittle in nature. It is typically used in compression, such as for columns, rather

Fig. 3.24   The fifth-century pillar at the Qutab Minar complex in Delhi, India is made of an extremely pure iron and remains virtually free from corrosion. (Photograph by John Stanley.)

than in tension. Cast iron cannot be forged or mechanically worked, but can be readily cast to form repetitive mouldings.

Steel is an alloy of iron and contains varying amounts of carbon and other elements according to the particular requirements of the end user. Mild steel, first produced in the mid-nineteenth century, has a carbon content of 0.25% or less, and has similar properties to wrought iron. Other forms of steel contain elements such as chromium, nickel and tungsten, and are used for their resistance to corrosion, decorative appearance or other specific characteristics. Weathering steel, such as Cor-Ten®, has the addition of copper (0.25–0.55%), which provides the desired tenacious oxide finish. Steels can be hot- and cold-worked or cast, depending on the composition of the metal and the use to which it is to be put.

Iron and steel construction

Cast and wrought iron were used extensively during the eighteenth and nineteenth centuries for industrial structures and the construction of framed and fire-proof buildings. The early use of cast iron, such as for the columns in St Anne's Church in Liverpool (1770–72), Abraham Darby's bridge at Coalbrookdale (1779) (Fig. 3.25), and Charles Bage's flax mill in Shrewsbury (1796–79), which enclosed both beams and columns in load-bearing masonry and had floors constructed of segmental brick vaults (jack arches), demonstrated both the qualities and the limitations of this material.

Later buildings, such as the Cooper Union Building in New York (1850) and the Chicago skyscrapers of the 1870s and 1880s, used wrought iron for individual members and structural frames in favour of cast iron. At the same time, buildings were being designed and built to make best use of the different structural qualities of the materials by combining cast and wrought iron in the frames and roofs, such as the Crystal Palace of 1850–51, or combinations of iron and timber for lightweight framed constructions.

Improvements in the production of steel, particularly with the intro-duction of the Bessemer converter (a method of mass-producing steel by blowing air through molten pig iron to remove impurities) in the 1850s, led to its growing use for building and engineering applications. Approval for the use of steel in the British construction industry was granted by the Board of Trade in 1877, and the first complete steel-framed building was erected in West Hartlepool in 1896. The construction of the Ritz Hotel in London (1904–6), which required the relaxation of the requirements of the London Building Act of 1894 for its riveted connections between beams and columns, marked the beginning of the popularity of steel for building construction, even though the building appears to be built of load-bearing

Fig. 3.25   Bridge at Coalbrookdale, Shropshire. (Photograph by Peter Swallow.)

masonry walls. Exposed steel frames, with lightweight cladding and glazing panels, became popular from the mid-twentieth century.

Non-ferrous metals

The use of non-ferrous metals in buildings is typically limited to the formation of sheet roof coverings, flashings, weatherings and plumbing (Fig. 3.26). Lead and copper have been used as traditional roof coverings in Britain since the Roman occupation, and also for the formation of pipework and waterproof containers (cisterns, sinks). Zinc appeared during the early nineteenth century as an alternative roof covering to lead (the earliest recorded use being for zinc tiles in 1832), and later as the basis for galvanising ferrous metals. Aluminium and its alloys have become widely available during the twentieth century, with typical uses being cladding, curtain wall and structural glazing systems, and internal finishes.

Fig. 3.26 Milled lead used on the lantern to the octagon at Ely Cathedral, Cambridgeshire.

Other metals are found in the construction and decoration of buildings, many being formed as an alloy of two or more parent metals. These have varying compositions according to the specific requirements of the application and user. Brass (an alloy of copper and zinc) and bronze (an alloy of copper and tin) are both used for casting small architectural elements, such as ironmongery, and for works of arts.

The precious metals of gold and silver may be present in the forms of decorative leaf (typically only 0.1 micrometre or four-millionths of an inch thick and laid over a prepared surface) and works of art. Pewter (an alloy of varying amounts of tin, copper, antimony and lead) has historically been used for the manufacture of domestic items, while tin has become familiar as a corrosion-resistant coating to iron and steel, and as a component in solder.

Modern metallic products

Steel has continued to be used in various forms for the construction and finishing of buildings, with different compositions and coatings used to impart colour and texture, and improve performance. The use of stainless steel with a coating of lead or terne (an alloy of lead and tin) demonstrates, for instance, the application of modern metals in the repair and mainte-nance of traditional buildings. Non-ferrous metals, particularly aluminium and its alloys, and often with a metallic or polymeric coating, are widely used for decoration and finishing, and for the formation of coordinated fixtures and fittings.

Glass

The term 'glass' describes an amorphous (non-crystalline), highly viscous, super-cooled liquid, which is made up principally of silicon-oxygen (SiO_2) units with an alkali flux of soda (sodium oxide) or potash (potassium oxide), and modifiers of lime (calcium oxide) or magnesium oxide. Various other chemical constituents may also be present, such as lead, boron, cobalt and gold, to impart specific characteristics or colours. Glass ceramics, by contrast, have a crystalline structure, which imparts high strength and shock resistance.

The earliest use of glass, for small ornaments found in the Near East, oc-curred in the third millennium BC, but it was not until the sixteenth century BC that glass-making had developed into a distinct industry (Freestone, 1991, p. 38). Evidence for window glass has been found at Roman sites, but coloured and painted glasses, principally for ecclesiastical use, were avail-able in Britain only from the eleventh century. Glass for general secular use was available from the seventeenth century.

Window glass has been made in a variety of ways, each giving the glass a distinct appearance. Early cast glass was rough finished, and had to be ground and polished to produce a clear glass. Crown glass was prominent from *c.* 1700 to the 1840s, and derives its sparkle from the manner in which it was spun from a blown globe. Cylinder glass, also known as 'broad' or 'muff' glass, was available from the mid-sixteenth century and became common in the following centuries. It was cut from a cylinder of blown glass, and has a smooth and less lively appearance to that of crown glass. Plate glass, which was made by pouring molten glass on to a metal table and rolling it flat, was developed in France in the seventeenth century and became the basis for glass production into the twentieth century. The float process, which supports the glass on a bath of molten tin during production, has become the principal method of manufacture since its introduction by Pilkington in 1959.

Modern glass and glass products

Modern sheet glasses may be wired, toughened, laminated, patterned (embossed, printed, etched, sandblasted, coloured), fire-resistant or surface-modified (low emissivity, tinted, reflective, coated), and used in single or multiple glazing systems. Specialist glasses include one-way observation glass, alarm glass, electromagnetic radiation-shielding glass, variable-transmission glasses (photochromic, thermochromic or electrochromic), and self-cleaning glass (Pilkington 'Avtiv').

Glass may also be used in a structural form, as glass blocks, as fibres in the manufacture of glass-fibre reinforced polyester (GRP), glass-fibre reinforced cement (GRC), glass-fibre reinforced gypsum (GRG) and glass-wool insulation, or foamed to make glass insulation products.

Bituminous products

Asphalt and bitumen are amorphous (non-crystalline) products that are typically resistant to the passage of water and water vapour. As such, they have long been used as waterproof coatings or coverings, but are liable to flow or creep under mechanical stress and soften in response to increases in temperature.

Asphalt is obtained naturally either in the form of 'rock' asphalt, which consists of a limestone impregnated with bitumen, or as 'lake' asphalt from sources such as Trinidad in the West Indies. Mastic asphalt describes a blended bitumen-based product that is heated and mixed with aggregates to give a product suitable for use as a covering to flat roofs. Bitumen is obtained from natural sources or as a by-product from the distillation of crude oil, and used in the liquid form or impregnated into other materials.

Other materials associated with asphalt and bitumen include coal tar, which is derived from the distillation of coal, and pitch, which is the residue resulting from the distillation of coal tar, wood tar or petroleum, and also from naturally occurring petroleum residues.

Modern bituminous products

Asphalt and bitumen are widely used for forming a waterproofing finish on roofs. Modern variations of the traditional mastic asphalt and built-up roofing systems include polymer-modified mastic asphalt, which enhances durability and flexibility, and high-performance and polymer-modified bitumen roofing felts, which improve flexibility and strength.

Modern materials

The use and development of materials for the construction and finishing of buildings since the mid-twentieth century has typically centred on improvements in manufacture and performance. Today's designers and specifiers have an enormous choice of materials, whether it be bricks and tiles or architectural fittings, and the age-old concern of transportation has been replaced by issues of energy efficiency, local distinctiveness and the revival of traditional crafts.

Perhaps the one material that has been developed and utilised in recent years, and achieved almost universal application, is plastic. Plastics, which can be shaped by applying heat or pressure, are typically formed from polymeric (having large molecules consisting of repeated units) synthetic resins, and can be defined as being either thermoplastic (may be repeatedly softened by heating and hardening again on cooling) or thermosetting (initially soft, but changed irreversibly to a hard rigid form on heating). In addition, elastomeric polymers have the ability to stretch and return to their original shape. Plastics contain various additives, such as pigments, plasticisers (increase flexibility), anti-oxidants (reduce oxidation), stabilisers (absorb ultraviolet light) and fillers (reduce costs and improve use), to modify their performance in a variety of applications (Fig. 3.27).

Plastics are formed in a variety of ways (extrusion, blowing, moulding, pressing) to suit their particular application. Some of the more common uses for plastics in the construction of buildings include:

- damp-proof courses and membranes (polythene – PE)
- vapour barriers (polythene – PE)
- roofing systems (polyvinyl chloride – PVC, ethylene propylene diene monomer – EPDM)
- rainwater goods (polyvinyl chloride – PVC)
- plumbing and sanitary goods (polypropylene – PP, polybutylene)
- drainage systems (polyvinyl chloride – PVC, acrylonitrile butadiene styrene – ABS)
- glazing and roof lights (polycarbonate – PC, polymethyl methacrylate – PMMA)
- architectural components (polyvinyl chloride – PVC, glass-reinforced polyester – GRP)
- movement joints (polytetrafluoroethylene – PTFE)
- electrical insulation (polyamide – PA)
- decorative components and finishes (phenol formaldehyde – PF, melamine formaldehyde – MF, urea formaldehyde – UF)
- bearings (rubber)

Fig. 3.27   Nelson's Monument, Great Yarmouth, Norfolk (1817–19). The figure of Britannia and the six caryatids around the open drum are constructed of glass-reinforced plastic and replaced original figures of Coade stone. The figures are formed in jointed sections over galvanised mild-steel armatures and support frames, with a concrete filling and etched surface finish.

- seals (neoprene)
- gaskets (neoprene, ethylene propylene diene monomer – EPDM)

Early plastics (such as cellulose nitrate, cellulose acetate, casein, and phenol formaldehyde or Bakelite) may also still be found in buildings, including characteristically brown Bakelite electrical fittings.

Building services

The servicing of buildings, and the ways in which these services are arranged and used, provides a fascinating means of understanding the lives

and expectations of those living in previous centuries. In an intriguing survey of domestic arrangements, Christina Hardyment (1997, pp. 11–27) contrasts three different properties and from this derives a distinction between what she calls *organic, rational* and *scientific* systems.

These systems are illustrated by Speke Hall in Liverpool, a timber-framed house of the sixteenth century that relied on servants for all the needs of the family and friends (*organic*); Florence Court in Fermanagh (Northern Ireland), designed and built in the mid-eighteenth century to ensure that the servants and their practical tasks were kept out of sight of the 'polite parts' of the house (*rational*); and Cragside in Northumberland, a Victorian mansion designed mainly by R. Norman Shaw that embraced technology (it was the first house in the world to be lit by hydroelectricity) and did away with the need for numerous servants and service rooms (*scientific*).

The National Trust 'Country House Technology Survey' was started in 1995 to record and document evidence of surviving technologies in its properties. The survey is ongoing, and has revealed that the Trust is the owner of some of the most important examples of early technology, including electrical and acetylene lighting, heating, internal transport and communications systems, water provision and sewage disposal.

The provision of services in today's buildings has become a subject of critical and growing complexity, and their presence has clearly altered the ways in which we choose to live and work. This move away from natural provision to the dominance of artificial services has also had a dramatic effect on how buildings are designed and built (Fig. 3.28). As one commentator (John Winter) has stated, 'making old buildings habitable, with electric light, hot and cold running water and central heating but without servants, will in the future be seen as one of the main achievements of twentieth-century building technology' (Davies, 1985, p. 39).

The complexity and variability of building services do not permit anything more in this chapter than a brief summary of what is likely to be found or required in buildings under survey (Table 3.5). Instead, it is perhaps of greater value to consider how such services are integrated into the building and what effect their presence might have on its condition and the comfort of its occupants.

Services for buildings and people

Modern buildings are typically designed around the services that will be provided for their users. Thus, lifts and elevators allow the rapid vertical transfer of people from one floor to another, and have been instrumental in the design of taller buildings. By contrast, advances in electronics and

Fig. 3.28   Brokaw-McDougall House, Tallahassee, Florida. Built in *c.* 1856–60, this house was designed to provide a comfortable internal environment with its veranda, shuttered windows, overhanging eaves, cupula and tall ceilings. The internal arrangement, with shaded windows and the cupola positioned over the stairway, provided ventilation throughout the house before the advent of air conditioning.

telecommunications have brought about a reappraisal of the role of people inside buildings, and the tall building has often become more a status symbol than an absolute necessity.

The introduction of today's services into yesterday's buildings does not, however, offer such freedoms, and the demands and expectations of clients have to be accommodated within the structural and material constraints imposed by the building and relevant limiting legislation (Fig. 3.29).

Where new services are planned and introduced, care has to be taken to ensure that what is required by one occupant or operation does not have a detrimental effect on another. It is equally important to consider how, for instance, increased heating or reduced natural ventilation levels will affect the performance of the building as a whole and of its individual elements or components. The reliance on instant forms of heat as opposed to background heating that makes best use of the thermal mass of a building can, for example, be seen as both wasteful and potentially damaging, whilst reduced levels of ventilation can be responsible for mould growth and the outbreak of fungal decay.

Table 3.5 Summary of building services.

• Water supply
    ○ treatment
    ○ cold water
    ○ hot water
	• Drainage and waste disposal
	    ○ foul water
	    ○ surface water
	    ○ grey water

• Heating
    ○ wet systems
    ○ dry systems
	• Lighting
	    ○ general background lighting
	    ○ specific task lighting

• Ventilation
    ○ natural
    ○ mechanical
	• Communications
	    ○ telecommunications
	    ○ computer networks

• Building management system (BMS)	• Environmental monitoring and control

• Mechanical conveyance
    ○ goods
    ○ personnel
	• Fire
	    ○ detection (heat and smoke)
	    ○ alarm
	    ○ fire-fighting (sprinklers)

• Lightning protection	

• Security
    ○ intruder detection
    ○ alarm
	• Domestic arrangements
	    ○ cooking
	    ○ washing

The servicing of historic buildings, which may be used for a variety of purposes and which often contain original finishes and fittings that are particularly susceptible to fluctuating environmental conditions, requires specific attention (Fig. 3.30). The approaches taken within buildings open to the public, such as National Trust properties (Sandwith & Stainton, 1991), museums and galleries (Thomson, 1986; Cassar, 1995; Staniforth, 1995; Lawson-Smith, 1998), is in this respect particularly worthy of note, and specialised advice should normally be sought when considering the introduction or modification of services.

Modern service installations and the comfort standards that they are designed to provide may also have a potentially serious effect on the health and well-being of the building occupants and users. It is increasingly being demonstrated that levels and standards of heat, light and ventilation, and the operation of air-conditioning systems, are responsible for significant levels of stress and discomfort, and that the reliance on artificially serviced or 'tight' buildings has been a cause of various building-related illnesses (see Chapter 4).

Greater concern is also now being expressed about the indoor environment in which people live and work. The number of chemicals used in the manufacture of building materials and internal finishes, and the increasing

Fig. 3.29   The changing needs of building users, such as in Hong Kong, often result in the addition or modification of building services to the detriment of the building.

amounts of other indoor pollutants present within the home or workplace, are clearly having an effect on health and productivity. The role of building services in meeting the demands of today's society, and the need to understand better the implications of often well-intentioned alterations or improvements on the performance of a building, are issues that require particular attention.

Building services of the future will no doubt be more closely integrated with the needs of both the building and its occupants, and may have a degree of intelligence that will allow the building to automatically respond to changing conditions within a range of predetermined responses (Fig. 3.31). This is already being partly addressed with products such as remote-controlled lighting, people-detectors and automatic sun-shades. In order to be truly intelligent, however, all the various systems within a building (such as heating, lighting and communications) must be integrated and

Fig. 3.30 The Gurney coke-fired stoves in Ely Cathedral, Cambridgeshire were converted to gas in the 1970s.

controlled by a single management system. This may include the wider use of motorised reflecting and shading devices, photovoltaic cells, wind turbines, intelligent lighting that responds to natural light levels and levels of occupancy, automatic devices (such as retractable roofs and motorised windows) to maximise the use of natural ventilation, and enhanced thermal insulation.

The building as a whole

Having acknowledged the variety and importance of materials and services in the study of buildings, it is the purpose of this section to illustrate how this information might be used to better understand how a building works and to consider some of the complex interactions between the building, its occupants and the environment.

Fig. 3.31   Queen's Building, De Montfort University, Leicester. The design strategy for this low-energy, naturally ventilated building allows it to function without the use of air conditioning, despite high internal heat gains in lecture theatres and computer laboratories. Ventilation for the auditoria relies on the stack effect of warm air rising and being replaced by cold air at low level.

The idea that a building is made up of various layers (see Chapter 2) is again useful in considering how changes or interventions have influenced the way it performs and responds to present-day demands. These 'layers' may represent successive phases of building, rebuilding or alteration in response to the changing needs; significant events in the history of the building or its owners (fire, abandonment); previous repairs; and works to conform to contemporary standards. Each will leave evidence, whether obvious or discreet, that has to be discovered and understood by those responsible for the well-being of the building, its occupants and contents (Fig. 3.32). It is also their responsibility to consider how buildings need to be further altered with additional and often increasing complex 'layers'. This will be considered later in Chapter 7.

The interpretation of these various layers can be performed at many levels, depending on the complexity of the building and the purpose for which it is being examined. It may be appropriate, for instance, to consider the architectural, archaeological or historical significance of the building in order to inform those concerned with the history of its art and design qualities (Clark, 2001). This, in turn, may be of value to those dealing

Fig. 3.32 Powered by compressed air, a pump was employed to fill the voids in the masonry corework at Lincoln Cathedral with Portland cement grout as part of the Special Repair Programme (1922–32). (Lincoln Cathedral Works Archive.)

with its repair and conservation, such as through preparation of a conservation management plan, and later for those responsible for its marketing and disposal (Fig. 3.33). A building may also be interpreted at a structural, material or environmental level as part of an investigation to assess performance or provide a diagnosis for a particular defect or failing. Examples of this might include identifying and recording deleterious materials or previous repairs that are now found to be inappropriate or damaging.

Such detailed investigation of the various layers of a building is, however, rarely performed unless required by some specific demand for information or in response to an identified problem. Those undertaking building

Fig. 3.33   CAD-based modelling and video animation used to record and interpret Denver cornmill, Norfolk in advance of proposed restoration. (Courtesy of Rob Ashton/UKgeomatics.)

surveys generally follow a particular prescribed format, depending on the nature of the building or needs of the client, and run the risk of seeing a building as a simple collection of components or parts, rather than as a complex three-dimensional assembly combining various layers of information.

Understanding buildings and building performance

The assessment of buildings has typically lacked accurate and comparable information, with much that is learnt being entrusted to memory or office files. As the need for more detailed information increases, fuelled either by the wishes of the client or by the corresponding labours of the professional adviser, it is important that the resulting documentation (whether it be a survey drawing or written report) be both accurate and appropriate.

Prescribed levels of assessment, whilst relevant for certain forms of construction or building type, are often unsuited for buildings that are old or of

unusual construction. With these, the nature and extent of the assessment will require specific knowledge and experience to ensure competence in the advice given. The fourth edition of the *RICS Appraisal and Valuation Manual* ('Red Book') requires valuers, for instance, to seek competent advice from a person with 'appropriate specialist knowledge' when dealing with properties of 'architectural or historic interest, or listed as such; or in a conservation area; or of unusual construction', unless they themselves are competent to give advice that would not be 'detrimental to the property's architectural or historic integrity, its future structural condition or conservation of the building fabric' (Royal Institution of Chartered Surveyors, 1995, 2003).

The practice of building pathology, as with that of 'forensic conservation' (Weaver, 1995, p. 30), demands a broader and often more detailed assessment with a typically greater reliance on scientific and technological methods. This need for proven qualitative and quantitative data is perhaps one of the clearest distinctions between the disciplines of building surveying and building pathology.

To illustrate how such an approach might be taken to develop an understanding for a particular building, three case studies have been chosen as representing differing forms, uses and periods of history. All are located at St Fagans: National History Museum, Cardiff; two have been moved from their original locations and re-erected at the Museum. There is public access into the buildings, and both Nant Wallter Cottage and the Oakdale Workmen's Institute are suitably interpreted and presented for the purposes of the Museum.

In each case study a simple checklist has been used to collect and present the information, and to make comparisons between how the buildings were constructed, serviced and used.

Case Study – Nant Wallter Cottage

General details

- Name of property: Nant Wallter Cottage
- Original location: Taliaris, Carmarthenshire
- Date of erection: *c.* 1770 (re-erected and opened to public in 1993)
- Original uses: Dwelling for poor estate worker and family
- Form of construction: Timber crucks and load-bearing clom (mix of clay, straw and stone dust) on stone plinth
- Number of storeys: One with half-loft
- Gross floor area: Ground floor 37 m^2; half-loft 11 m^2

Materials audit

- Foundations: Originally stone; as rebuilt, concrete strip
- External walls: 750 mm thick clom on stone plinth with limewash finish
- Roof structure and covering: Two pairs of scarfed crucks, with wattle, gorse and straw thatch covering
- Internal walls and partitions: Timber posts with wattle and clay daub
- Floors: Beaten earth (clay and lime)
- Ceilings: Not applicable
- Doors: Oak boards
- Windows: Small-paned, non-opening, to bedroom and living area
- Stairs: Not applicable
- Internal finishes: Limewash
- Fixtures and fittings: Re-created late eighteenth-century interior
- Site and environment: Stone-flagged path, thorn hedge
- Summary of principal materials: Clay, straw (clom and thatch), stone, stone dust, timber, lime, gorse, glass

Services audit

- Heating: Peat fire with wattled chimney canopy above fireplace
- Lighting: Natural from windows and door
- Ventilation: Natural
- Water supply: Originally natural spring (Nant Wallter translates as Wallter's brook or stream)
- Cooking: Open fire
- Drainage and waste disposal: None
- Communications: None
- Mechanical conveyance: None

Performance

- Thermal transmittance (U-value): External clom wall construction, say 0.75 W/m^{2}K
- Glazing: 0.38 m^2 (1.8% front elevation)
- Original pattern of usage: Dwelling for a family of probably four or five persons
- Present pattern of usage: Estimated 150,000 persons per annum (based on 40% of visitors to Museum)

Additional comments

- Clom laid in 100–200 mm layers, each allowed to dry before next laid on top
- Base plinth wall used 50 tons of stone
- Walls used approx. 46 tons each of clay, stone dust and coarse sand, with a total weight of 136 tons of clom (46m^3); each ton of clom contained equivalent of one bale of wheat straw
- In addition to two pairs of scarfed oak crucks and purlin poles, approx. 300 split poles (oak and ash) used from wall head to first purlin, and further 100 split poles (200–300 mm spacing) laid between purlin and ridge; approx. 800 × 3 m long hazel rods (25–50 mm diameter) woven between letter poles to form base for underthatch
- Half acre of densely-packed gorse cropped and pruned to remove excess wood; laid to depth of 300 mm as underthatch; on top laid matting of straw, equivalent to approx. 20 bales, laid to depth of 250 mm; thatch of wheat straw formed on top
- Cottage took 2–3 men about one year to complete
- For further information on the use of earth construction in Wales, see: Nash (1995).

(a) Appearance of the cottage.

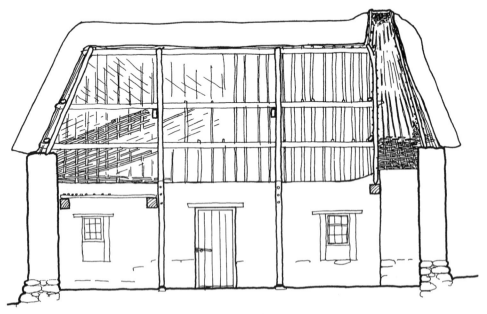

(b) Cross-sectional drawing indicating construction.

Case Study – Oakdale Workmen's Institute

General details

- Name of property: Oakdale Workmen's Institute
- Original location: Oakdale, Gwent
- Date of erection: 1916 (re-erected and opened to public in 1995)
- Original uses: assembly and entertainment (library, reading rooms, auditorium)
- Form of construction: Load-bearing masonry
- Number of storeys: Two
- Gross floor area: Ground floor 266.7 m^2; first floor 237.6 m^2

Materials audit

- Foundations: Concrete strip
- External walls: 480 mm thick coursed Penant sandstone, with facade and details in Portland stone
- Roof structure and covering: Softwood trusses and common rafters. Clay tile covering hung on battens. Lead flashings and weatherings. Central ventilator
- Internal walls and partitions: Common brickwork with plastered surfaces
- Floors: Sandstone and terrazzo step; ground-floor rooms herringbone wood block; first-floor auditorium wood strip (sprung floor) supported by steel joists
- Ceilings: Plaster
- Doors: Brass door furniture
- Windows: Plain and stained glass
- Stairs: Concrete
- Internal finishes: Glazed ceramic tiles; paints, stains and varnishes
- Fixtures and fittings: Fabric curtains, drapes (auditorium) and shelving
- Site and environment: Tarmac and gravel surfacing; stone boundary wall with wrought-iron railings
- Summary of principal materials: Stone, lime, cement, concrete, timber, fired clay, terrazzo, steel, iron (cast and wrought), lead, copper, zinc (galvanising)

Services audit

- Heating: Originally coal-fired central heating; as rebuilt, oil-fired central heating
- Lighting: Electric (originally generated for coal mines)
- Ventilation: Natural, with ventilator to first-floor auditorium
- Water supply: Mains
- Cooking: Not applicable
- Drainage and waste disposal: Foul and surface water to main drains
- Communications: Telephone
- Mechanical conveyance: As rebuilt, disabled lift

Performance

- Thermal transmittance (U-value): External stone wall construction, say 1.56 W/m^{2}K
- Glazing: 11.1 m^2 (8.6% front elevation)
- Original pattern of usage: Partly in use at all times
- Present pattern of usage: Estimated 200,000–250,000 persons per annum (based on 50–65% of visitors to Museum)

Additional comments

- First-floor auditorium was used for concerts, public meetings, dances, lectures and events, and also served as a cinema
- The auditorium could originally accommodate seating for up to 250 people; this is now restricted to 180 people
- The Institute's minute books record that in the late 1930s the interior was repainted in a 'marbled effect'; this colour scheme has been reproduced in the re-erected building
- For further information on workmen's institutes, and the Oakdale Institute in particular, see: Nash *et al.* (1995).

(a) Appearance of the building.

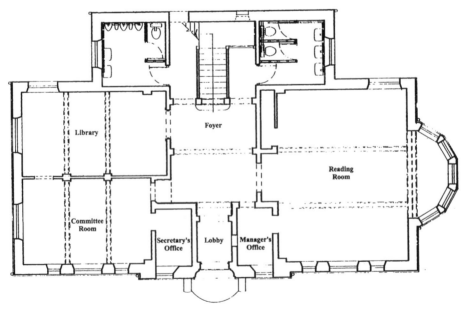

(b) Ground floor plan.

Case Study – Entrance Hall to Visitors' Centre at St Fagans

General details

- Name of property: Entrance hall (ground floor only) to visitors' centre
- Original location: Purpose-built in present location
- Date of erection: 1972 (altered since)
- Original uses: Designed for enquiries desk and cloaks area, with separate external access for shop to side of main entrance; now with visitor reception, ticket desk and shop
- Form of construction: Steel frame with cavity walling
- Number of storeys: One and two storey
- Gross floor area: Ground floor approx. 305 m^2

Materials audit

- Foundations: Concrete
- External walls: Quartz-concrete brick external leaf, cavity, internal blockwork leaf
- Roof structure and covering: Flat roof with built-up roof deck
- Internal walls and partitions: Quartz-concrete brick
- Floors: Clay tile
- Ceilings: Plasterboard and skim
- Doors: Original single-glazed steel-framed doors and windows to front and inner courtyard replaced *c.* 1996 with double-glazed aluminium units
- Windows: Single-glazed timber units
- Stairs: Concrete with clay tile treads
- Internal finishes: Paint
- Fixtures and fittings: Planters to side windows by desk
- Site and environment: Paved entrance beneath tented canopy with steps and ramp from car park; grass and hedging
- Summary of principal materials: Steel, concrete, brick, block, plaster, timber, glass

Services audit

- Heating: Gas-fired central heating
- Lighting: Electric supplemented by photovoltaic panels providing energy for lighting to concourse and steps
- Ventilation: Natural
- Water supply: Mains
- Cooking: Restaurant facilities
- Drainage and waste disposal: Foul and surface water to main drains
- Communications: Electronic and telecommunications
- Mechanical conveyance: None

Performance

- Thermal transmittance (U-value): External cavity wall construction, say 0.96 W/m^{2}K
- Glazing: 48.6 m^2 (82% front elevation)
- Original pattern of usage: Designed and used as entrance hall to visitors' centre
- Present pattern of usage: Estimated 360,000–400,000 persons per annum

Additional comments

- Building designed as a modular (9 ft) structure using steel and reinforced concrete beams
- Oil was not apparently considered as a fuel for heating the building, with gas-fired heating instead
- Flat roof to upper landing level has proved problematic due to rainwater penetration

(a) Entrance to visitors' centre.

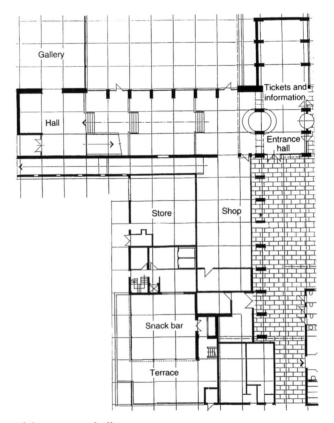

(b) Plan of the entrance hall.

Assessing building performance

Although assessments, such as those given in the above three case studies, bring together the various layers that make up a particular building and allow it to be better understood, there are methods or models that may be used to understand more fully the actual performance of a building.

The *Building Performance Model*, developed by the Building Performance Research Unit at Strathclyde University in the 1960s, is one such model that seeks to define the relationships between people and the building environment (Building Performance Research Unit, 1972). The model is based around five systems – building, environmental, activity, objectives and resources – and provides a means by which the relative importance of each system and subsystem can be identified (albeit open to influences such as site, location, climate, culture, society, politics, economics and business needs) and considered in the design of new buildings or the alteration and adaptation of old (Fig. 3.34).

More recently, the use of post-occupancy evaluations (POEs) has focused on building occupants and their particular needs, and has attempted to match buildings more closely with their users. Such an evaluation might include (Preiser *et al.*, 1988):

- technical elements (e.g. fire safety, structural integrity, sanitation, durability, acoustics, lighting)
- functional elements (e.g. operational efficiency, productivity, work flow, organisation)
- behavioural elements (e.g. privacy, symbolism, social interaction, density, territoriality)

Such work has demonstrated the importance and current undervaluing of design briefing, and the problems of redundancy for buildings of complex design. Achieving a 'loose fit' for a building, which provides flexibility in internal layout and usage, seems to offer the key for future designs and adaptations, and brings us a little closer to achieving a 'universally usable building' (Kernohan & Wrightson, 1997).

Performance measurement, together with sustained improvements in quality and efficiency, and reductions in construction costs, time and defects, have been identified as key components for improvement and change within the British construction industry (DETR, 1998). *Rethinking Construction*'s project-based Environmental Performance Indicators (EPIs) provide benchmarks against which designers and users of buildings can compare performance and set targets, whereas the primary purpose of Construction Related Sustainability Performance Indicators is to 'commence the

practical task of implementing a sustainable approach to the future development and modification of the "built environment", while also playing part in ensuring a flourishing future for the "natural environment" by carrying out sufficient repair to past, present and potential future damage directly or indirectly caused by construction' (IT Construction Forum, 2006).

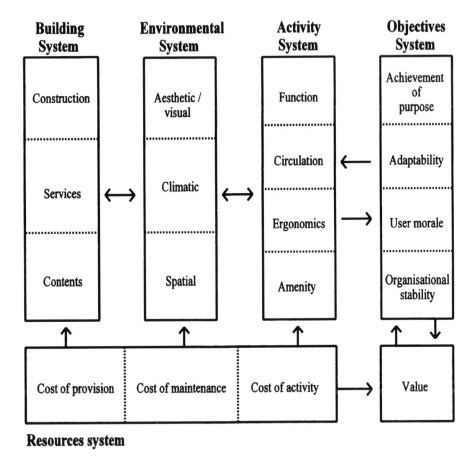

Fig. 3.34   Conceptual performance model of the system of building and people.

References

Addleson, L. (1972) *Materials for Buildings – Volume 1: Physical and Chemical Aspects of Matter and Strength of Materials; Volume 2 – Water and its Effects (1); Volume 3: Water and its Effects (2).* London: Iliffe Books.

Beard, G. (1981) *Craftsmen and Interior Decoration in England 1660–1820.* London: Bloomsbury Books.

Boniface, S. (1998) Dorset leads the way in new-build thatch. *Chartered Surveyor Monthly,* **7** (10), 39–41.

British Standards Institution (1993) *Glossary of Terms used in Terotechnology*. 4th ed. BS 3811. London: BSI.

Building Performance Research Unit (1972) *Building Performance*. Barking: Applied Science Publishers.

Cassar, M. (1995) *Environmental Management*. London: Routledge.

Clark, K. (2001) Informed *Conservation: Understanding Historic Buildings and their Landscapes for Conservation*. London: English Heritage.

Clifton-Taylor, A. (1987) *The Pattern of English Building*. 4th ed. London: Faber & Faber.

Davies, C. (1985) Conservative views. *Architects' Journal – Renovation*, 21–28 August.

Dean, Y. (1996a) *Mitchell's Finishes*. 4th ed. Harlow: Addison Wesley Longman.

Dean, Y. (1996b) *Mitchell's Materials Technology*. Harlow: Addison Wesley Longman.

DETR (1998) *Rethinking Construction: Report of the Construction Task Force on the Scope of Improving the Quality and Efficiency of UK Construction*. London: DETR.

Everett, A. & Barritt, C. (1994) *Mitchell's Materials*. 5th ed. Harlow: Addison Wesley Longman.

Freestone, I. (1991) Looking into Glass. In: *Science and the Past* (ed. S. Bowman), pp. 37–56. London: British Museum Press.

Hall, D. (1998) Poppyhead: The recreation of terracotta poppyheaded pew ends at the Church of St Stephen and All Martyrs. *Building Conservation Directory: The Conservation and Repair of Ecclesiastical Buildings*, 5th edn. Tisbury: Cathedral Communications, pp. 18–19.

Hardyment, C. (1997) *Behind the Scenes: Domestic Arrangements in Historic Houses*. 2nd ed. London: National Trust Enterprises.

Heyman, J. (1997) *The Stone Skeleton: Structural Engineering of Masonry Architecture*. Cambridge: Cambridge University Press.

Hudson, J.D. & Sutherland, D.S. (1990) In: *Stone: Quarrying and Building in England AD 43–1525* (D. Parsons ed), pp. 16–32. London: Phillimore.

IT Construction Forum. (2006) Performance Indicators, www.itconstructionforum.org.uk/itsustainability, accessed 8 May 2006.

Kernohan, D. & Wrightson, B. (1997) Universally useable buildings – A manager's guide. *International Journal of Facilities Management*, **1** (1), 41–9.

Lawson-Smith, P. (1998) Environmental control in historic buildings. *Journal of Architectural Conservation*, **4** (1), 42–55.

Lyons, A.R. (1997) *Materials for Architects and Builders: An Introduction*. London: Arnold.

Middleton, A. (1991) Ceramics: materials for all reasons. In: *Science and the Past* (S. Bowman ed), pp. 16–36. London: British Museum Press.

Nash, G.D. (1995) Earth-built structures in Wales. Presented for *Study and Conservation of Earth as a Building Material*, Institute of Advanced Architectural Studies, University of York.

Nash, G.D., Davies, A.J. & Thomas, B. (1995) *Workmen's Halls and Institutes: Oakdale Workmen's Institute*. Cardiff: National Museums & Galleries of Wales.

Norman, C. (1998) *Baddesley Clinton*. London: National Trust.

Preiser, W., Rabinowitz, Z. & White, E. (1988) *Post-Occupancy Evaluation*. New York: Van Nostrand Reinhold.

Rackham, O. (1986) *The History of the Countryside*. London: J.M. Dent.

Royal Institution of Chartered Surveyors (1995) *RICS Appraisal and Valuation Manual*, 4th ed. Annex A to Practice Statement 9. London: RICS.

Royal Institution of Chartered Surveyors (2003) *RICS Appraisal and Valuation Standards*, 5th ed. London: RICS.

Sandwith, H. & Stainton, S. (1991) *The National Trust Manual of Housekeeping*. 2nd ed. London: Penguin Books Ltd.

Schilderman, T. (1998) DIY house building in Zimbabwe. *Intermediate Technology – Small World*, Issue 25, p. 9.

Staniforth, S. (1995) Environmental control in historic buildings. *ASCHB Transactions*, **20**, 33–42.

Stone, E.C., Lindsley, D.H., Pigott, V., Harbottle, G. & Ford, M.T. (1998) From shifting silt to solid stone: The manufacture of synthetic basalt. *Science*, **280** (5372), 26 June, pp. 2091–93.

Tickell, O. (1998) A cut above: Star-cutting gives you more wood from the trees. *New Scientist*, **159** (2141), 4 July, p. 10.

Thomson, G. (1986) *The Museum Environment*. London: Butterworth.

Torraca, G. (1988) *Porous Building Materials: Materials Science for Architectural Conservation*. 3rd ed. Rome: ICCROM.

Weaver, M.E. (1995) Forensic conservation and other current developments in the conservation of heritage resources and the built environment. *Journal of Architectural Conservation*, **1** (3), November.

Further reading

Adams, E.C. (1983) *Science in Building for Craft and Technician Students*. 2nd ed. London: Hutchinson.

Airs, M. (1998) *The Tudor & Jacobean Country House: A Building History*. Godalming: Bramley Books.

Allen, W. (1997) *Envelope Design for Buildings*. London: Architectural Press.

Ashurst, J. & Ashurst, N. (1988) *Practical Building Conservation – Volume 1: Stone Masonry; Volume 2: Brick, Terracotta and earth; Volume 3: Mortars, Plasters and Renders; Volume 4: Metals; Volume 5: Wood, Glass and Resins*. Aldershot: Gower Technical Press.

Ashurst, J. & Dimes, F. (1998) *Conservation of Building and Decorative Stones*. 2nd ed. London: Butterworth-Heinemann.

Baer, N.S., Fitz, S. & Livingston, R.A. (eds.) (1998) *Conservation of Historic Brick Structures*. Shaftesbury: Donhead Publishing.

Baird, G., Gray, J., Kernohan, D., Issacs, N. & McIndoe, G. (1996) *Building Evaluation Techniques*. New York: McGraw-Hill.

Beckman, P. (1994) *Structural Aspects of Building Conservation*. London: McGraw-Hill.

Blanc, A. (1994) *Mitchell's Internal Components*. Harlow: Addison Wesley Longman.

Brunskill, R. (1987) *Illustrated Book of Vernacular Architecture*. 3rd ed. London: Faber & Faber.

Brunskill, R. (1994) *Timber Building in Britain*. 2nd ed. London: Victor Gollancz.

Brunskill, R. (1997) *Brick Building in Britain*. 2nd ed. London: Victor Gollancz.

Brunskill, R. (1997) *Houses and Cottages of Britain: Origins and Development of Traditional Materials*. London: Victor Gollancz.

Building Research Establishment (1997) *Selecting Natural Building Stones*. Digest 420. Garston: BRE.

Burberry, P. (1997) *Mitchell's Environment and Services*. 8th ed. Harlow: Addison Wesley Longman.

Caroe, A. & Caroe, M. (1984) *Stonework: Maintenance and Surface Repair*. London: CIO Publishing.

Charles, F.W.B. & Charles, M. (1995) *Conservation of Timber Buildings*. London: Donhead Publishing.

Clements-Croome, D. (2004) *Intelligent Buildings: Design Management and Operation*. London: Thomas Telford.

Clifton-Taylor, A. (1987) *The Pattern of English Building*. 4th ed. London: Faber & Faber.

Clifton-Taylor, A. & Ireson, A.S. (1993) *English Stone Buildings*. 2nd ed. London: Victor Gollancz Ltd.

Coombes, M.C. (1983) Working with Existing Buildings. *Architects' Journal*, 24 and 31 August, pp. 85–88.

Davey, N. (1961) *A History of Building Materials*. London: Phoenix House.

Douglas, J. (1993/4) Developments in appraising the total performance of buildings. *Structural Survey*, **12** (6), 10–15.

English Heritage (1998) *Stone Slate Roofing: Technical Advice Note*. London: English Heritage.

Feld, J. & Carper, K. (1997) *Construction Failure*. 2nd ed. New York: John Wiley.

Foster, M. (ed.) (1983) *The Principles of Architecture: Style, Structure and Design*. Oxford: Phaidon Books.

Gale, W., (1979) *Iron and Steel*. Museum booklet No. 20.04. Ironbridge: Ironbridge Gorge Museum Trust.

Gauld, B. (1988) *Structures for Architects*. 2nd ed. Harlow: Longman Scientific and Technical.

Geeson, A.G. (1953) *Building Science – Volume 1: For Students of Architecture and Building; Volume 2 – Materials; Volume 3 – Structures*. London: English Universities Press.

Gordon, J.E. (1976) *The New Science of Strong Materials or Why You Don't Fall Through the Floor*. 2nd ed. London: Penguin Books.

Gordon, J.E. (1978) *Structures or Why Things Don't Fall Down*. London: Penguin Books.

Hall, F. (1995) *Building Services and Equipment – Volumes 1, 2 and 3*. London: Longman Scientific and Technical Publications.

Hart, D. (1991) *The Building Slates of the British Isles*. Garston: BRE.

Hill, P. R. & David, J.C.E. (1995) *Practical Stone Masonry*. London: Donhead Publishing.

Hodges, H. (1989) *Artifacts: An Introduction to Early Materials and Technology*. 3rd ed. London: Gerald Duckworth.

Hollis, G. (1996) Manufacturing faience and terracotta. *Context*, No 52, pp. 11–14.

Holmes, S. & Wingate, M. (1997) *Building with Lime: A Practical Introduction*. London: IT Publications.

Hughes, P. (1986) *The Need for Old Buildings to "Breathe"*. Information Sheet 4, London: SPAB.

Induni, B. & Induni, L. (1990) *Using Lime*. Lydeard St Lawrence: Induni.

International Council on Monuments and Sites. (2003) *Principles for the Analysis, Conservation and Structural Restoration of Architectural Heritage*. Paris: ICOMOS.

Innocent, C.F. (1971) *The Development of English Building Construction*. Newton Abbot: David & Charles.

Ireson, A. (1987) *Masonry Conservation and Restoration*. Rhosgoch: Attic Books.

Leary, E. (1983) *The Building Limestones of the British Isles*. Garston: BRE.

Leary, E. (1986) *The Building Sandstones of the British Isles*. Garston: BRE.

Lynch, G. (1994) *Brickwork: History, Technology and Practice – Vols 1 and 2*. London: Donhead Publishing.

Macdonald, S. (2003) *Concrete: Building Pathology*. Oxford: Blackwell Science.

McEvoy, M. (1994) *Mitchell's External Components*. Harlow: Addison Wesley Longman.

McKay, W. (2005) *McKay's Building Construction*, Vols I–III (facsimile of 1938–44 edition). Shaftesbury: Donhead.

Melville, I.A. & Gordon, I.A. (1973) *The Repair and Maintenance of Houses*. London: Estates Gazette Limited.

Parsons, D. (ed.) (1990) *Stone: Quarrying and Building in England AD 43–1525*. London: Phillimore.

Pearson, G.T. (1992) *Conservation of Clay and Chalk Buildings*. London: Donhead Publishing.

Salmond, C. (1995) Period window glass: A brief history of glass. *Context*, No. 48, 6–8.

Seeley, I. (1993) *Building Technology*. 4th ed. London: Macmillan Press.

Smith, B.J. (1976) *Construction Science – Volumes 1 and 2*. London: Longman.

Stevenson, G. (2003) *Palaces for the People/Prefabs in Post-War Britian*. London: Batsford.

Stratton, M.J. (1993) *The Terracotta Revival*. London: Victor Gollancz.

Strike, J. (1991) *Construction into Design: The Influence of New Methods of Construction on Architectural Design 1690–1990*. Oxford: Butterworth-Heinemann.

Stroud-Foster, J. (1994) *Mitchell's Structure and Fabric: Part 1*. 5th ed. Harlow: Addison Wesley Longman.

Stroud-Foster, J. & Harington, R. (1994) *Mitchell's Structure and Fabric: Part 2*. 5th ed. Harlow: Addison Wesley Longman.

Sunley, J. & Bedding, B. (eds.) (1985) *Timber in Construction*. London: B.T. Batsford/TRADA.

Taylor, G. (1991) *Construction Materials*. London: Longman Scientific & Technical.

Teutonico, J.M. (ed.) (1996) *Architectural Ceramics: Their History, Manufacture and Conservation*. London: James & James (Science Publishers).

Weaver, M.E. (1993) *Conserving Buildings: A Guide to Techniques and Materials*. New York: John Wiley and Sons.

West, R.C. (1987) *Thatch: A Manual for Owners, Surveyors, Architects and Builders*. Newton Abbot: David and Charles.

Wright, A. (1991) *Craft Techniques for Traditional Buildings*. London: B.T. Batsford Ltd.

Chapter 4
Defects, Damage and Decay

What is a building defect?

A defect may be considered to be a failing or shortcoming in the function, performance, statutory or user requirements of a building, and might manifest itself within the structure, fabric, services or other facilities of the affected building. When an inspection or survey is being undertaken, the set of requirements for the particular building type or use will help to set performance benchmarks against which the building can be measured. Where a performance benchmark is not achieved, this indicates a defect or deficiency, the severity of which is gauged by reference to the benchmark.

The severity of a building defect and the associated levels of damage, deterioration or decay currently present or expected to affect the building and its occupants are similarly related to the perceptions and expectations of the owner and occupier, and to various other stakeholders with interests in the well-being of the property. The defect, or the action required to reduce or remove its effect on the building, will typically be ranked according to a predetermined set of priorities for repair, maintenance or other works to improve either performance or capability.

Nature of building defects

The various elements and associated service installations that make up a building, together with the contents that allow it to be used and enjoyed, are susceptible to various forms of defect or fault. Past and present research has helped to identify the principal causes, yet many of the problems relating to poor-quality design, construction, repair and maintenance continue to reduce the utility and value of the nation's existing building stock.

Classification of agencies or mechanisms that adversely affect buildings is complex, for much depends on the particular conditions in and around the building, and the uses to which it is put. The classification proposed in ISO 6241:1984 offers a useful distinction between those agents acting outside and inside buildings (Table 4.1).

Table 4.1 Classification of agents acting outside and inside buildings.

Agents	Acting outside the building		Acting inside the building	
	Atmosphere	Ground	Occupancy	Design consequences
Mechanical agents				
Gravitation	Snow and rainwater loads	Ground and water pressure	Live loads	Dead loads
Forces and imposed or restrained deformations	Ice formation pressure, thermal and moisture expansion	Subsidence, slip	Handling forces, indentation	Shrinkage, creep, forces and imposed deformations
Kinetic energy	Wind, hail, external impacts, sand-storm	Earthquakes	Internal impacts, wear	Water hammer
Vibration and noises	Wind, thunder, aeroplanes, explosions, traffic, machinery noises	Traffic and machinery vibrations	Noise and vibration from music, dancers, domestic appliances	Services noises and vibrations
Electromagnetic agents				
Radiation	Solar radiation, radioactive radiation	Radioactive radiation	Lamps, radioactive radiation	Radiating surfaces
Electricity	Lightning	Stray currents	–	Static electricity, electrical supply
Magnetism	–	–	Magnetic fields	Magnetic fields
Thermal agents	Heat, frost, thermal shock	Ground heat, frost	User-emitted heat, cigarette	Heating, fire

Contd.

Table 4.1 Contd.

Agents	Acting outside the building		Acting inside the building	
	Atmosphere	Ground	Occupancy	Design consequences
Chemical agents				
Water and solvents	Air humidity, condensations, precipitations	Surface and ground water	Water sprays, condensation, detergents, alcohol	Water supply, waste water seepage
Oxidising agents	Oxygen, ozone, nitrous oxides	Positive electrochemical potentials	Disinfectant, bleach	Positive electrochemical potentials
Reducing agents	–	Sulphides	Agents of combustion, ammonia	Agents of combustion, negative electrochemical potentials
Acids	Carbonic acid, bird droppings, sulphuric acid	Carbonic acid, humic acids	Vinegar, citric acid, carbonic acid	Sulphuric acid, carbonic acid
Bases	–	Lime	Sodium, potassium, ammonium hydroxides	Sodium hydroxide, cement
Salts	Salty fog	Nitrates, phosphates, chlorides, sulphates	Sodium chloride	Calcium chloride, sulphates, plaster
Chemically neutral	Neutral dust	Limestone, silica	Fat, oil, ink, neutral dust	Fat, oil, neutral dust
Biological agents				
Vegetable and microbial	Bacteria, seeds	Bacteria, moulds, fungi, roots	Bacteria, house plants	–
Animal	Insects, birds	Rodents, termites, worms	Domestic animals	–

Research into the technical quality of the design and construction of new housing some years ago revealed 955 different kinds of faults in low-rise, mainly two-storey, housing (Bonshor & Harrison, 1982). Just under half of these were judged to have originated on site, slightly fewer in design, and with a small remainder related to materials and components. Other significant factors included inadequate design information and poor site practices.

The detailed results of this research (summarised in BRE, 1988) provide a useful attribution of (a) fault type to particular building elements and (b) the effects on building performance (Fig. 4.1). More recent research,

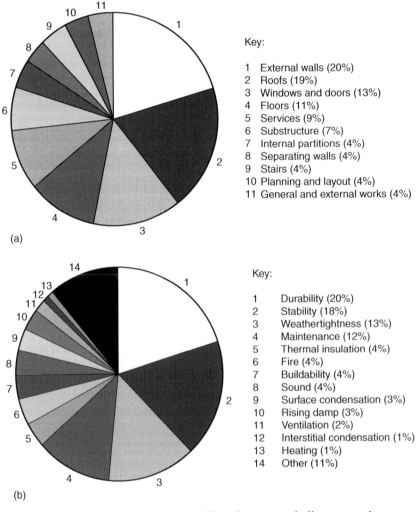

Key:

1 External walls (20%)
2 Roofs (19%)
3 Windows and doors (13%)
4 Floors (11%)
5 Services (9%)
6 Substructure (7%)
7 Internal partitions (4%)
8 Separating walls (4%)
9 Stairs (4%)
10 Planning and layout (4%)
11 General and external works (4%)

(a)

Key:

1   Durability (20%)
2   Stability (18%)
3   Weathertightness (13%)
4   Maintenance (12%)
5   Thermal insulation (4%)
6   Fire (4%)
7   Buildability (4%)
8   Sound (4%)
9   Surface condensation (3%)
10 Rising damp (3%)
11 Ventilation (2%)
12 Interstitial condensation (1%)
13 Heating (1%)
14 Other (11%)

(b)

Fig. 4.1 Attribution of fault type to building elements and effects on performance.

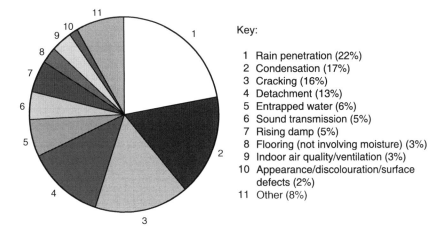

Key:

1 Rain penetration (22%)
2 Condensation (17%)
3 Cracking (16%)
4 Detachment (13%)
5 Entrapped water (6%)
6 Sound transmission (5%)
7 Rising damp (5%)
8 Flooring (not involving moisture) (3%)
9 Indoor air quality/ventilation (3%)
10 Appearance/discolouration/surface
    defects (2%)
11 Other (8%)

Fig. 4.2 Attribution of defect types.

conducted by the BRE Advisory Service and based on its database of building defects (Trotman, 1994), identified the ten main types of defect (Fig. 4.2). Dampness, in the form of rain penetration, condensation, entrapped moisture and rising damp, accounts for a high proportion of the categories. The Construction Quality Forum (CQF), which operated a defects database based on member responses, also analysed (a) the elements or locations where defects occur and (b) how defects contribute to poor building performance (Fig. 4.3). This was based on 862 database entries, 303 of which related to residential and 559 to non-residential construction (Construction Quality Forum, 1997).

Housing Association Property Mutual similarly analysed data from audits of 2120 schemes (about 31,000 dwellings) and identified common potential defects for six particular elements of construction (foundations, ground floors, external masonry walls, pitched roofs, separating walls and intermediate floors) (HAPM, 1997). A general overview of this work indicated particular issues relating to non-compliance with Buildings Regulations, lack of technical guidance, conflicting requirements and utilisation of marginal sites. The work also confirmed that the majority of defects occur through failure to achieve adequate standards with traditional forms of construction, rather than with novel or innovative construction.

Research carried out on behalf of Maintain our Heritage (Watt & Colston, 2003) regarding the maintenance of heritage properties has provided an analysis of (a) building elements or components affected by defects and (b) the frequency of occurrence for building defects (Fig. 4.4). It is

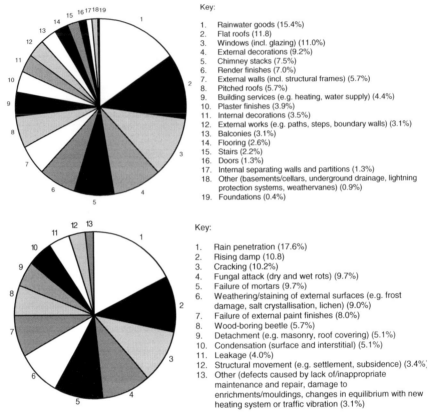

Key:

1. Rainwater goods (15.4%)
2. Flat roofs (11.8)
3. Windows (incl. glazing) (11.0%)
4. External decorations (9.2%)
5. Chimney stacks (7.5%)
6. Render finishes (7.0%)
7. External walls (incl. structural frames) (5.7%)
8. Pitched roofs (5.7%)
9. Building services (e.g. heating, water supply) (4.4%)
10. Plaster finishes (3.9%)
11. Internal decorations (3.5%)
12. External works (e.g. paths, steps, boundary walls) (3.1%)
13. Balconies (3.1%)
14. Flooring (2.6%)
15. Stairs (2.2%)
16. Doors (1.3%)
17. Internal separating walls and partitions (1.3%)
18. Other (basements/cellars, underground drainage, lightning protection systems, weathervanes) (0.9%)
19. Foundations (0.4%)

Key:

1. Rain penetration (17.6%)
2. Rising damp (10.8)
3. Cracking (10.2%)
4. Fungal attack (dry and wet rots) (9.7%)
5. Failure of mortars (9.7%)
6. Weathering/staining of external surfaces (e.g. frost damage, salt crystallisation, lichen) (9.0%)
7. Failure of external paint finishes (8.0%)
8. Wood-boring beetle (5.7%)
9. Detachment (e.g. masonry, roof covering) (5.1%)
10. Condensation (surface and interstitial) (5.1%)
11. Leakage (4.0%)
12. Structural movement (e.g. settlement, subsidence) (3.4%)
13. Other (defects caused by lack of/inappropriate maintenance and repair, damage to enrichments/mouldings, changes in equilibrium with new heating system or traffic vibration (3.1%)

Fig. 4.3 Attribution of defects to building elements and corresponding lack of performance.

evident that excess moisture is a common problem whatever the age of the building.

Causes and effects of defects, damage and decay

Having considered the origins of building defects, it is important to correctly identify and assess the particular agency or mechanism that is affecting the building before attempting to undertake remedial works. The purpose of this chapter is therefore to consider and explain the principal causes of defects, damage and decay.

The effects and consequences of these agencies or mechanisms will vary, depending on the construction, location, use and condition of the affected building. The corresponding risks to the well-being of the building, which need also to be considered as part of a survey or inspection of a building,

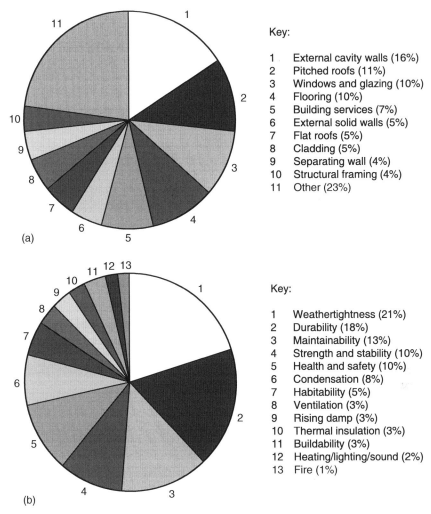

Fig. 4.4   Building elements/components affected by defects and corresponding frequency of defect.

have usefully been proposed by Camuffo (1997, pp. 84–5) and form the basis for Table 4.2.

Atmospheric and climatic action

The climate of the British Isles is a product of both global weather patterns and regional variations such as latitude, topography (mountains, moorland) and maritime influences. Weather elements (rainfall, temperature,

Table 4.2 Natural (a) and anthropogenic (b) risk factors affecting buildings.

(a) Natural risk factors	Nature	Timing	Scale	Probability
Biological factors Animals, insects, fungi, plants	biological	long-term	local/regional	continuous
Meteorological factors Solar radiation, precipitation, frost	physical	long-term	local	continuous cycles
Air humidity and temperature	physical	long-term	local	continuous cycles
Lightning	physical	rapid onset	local	infrequent
Gale force winds, tornadoes, hurricanes	physical	rapid onset	regional	rare/ infrequent
Geochemical hazards Groundwater erosion	physicochemical	long-term	local	continuous
Soluble salts	physicochemical	long-term	local	continuous
Natural hazards Soil movement	physical	long-term	regional	continuous
Sea surge	physical	rapid onset	local	frequent
River and sea floods	physical	rapid onset	local/regional	rare/ infrequent
Earthquakes	physical	rapid onset	local/regional	rare/ infrequent
Fires	physical	rapid onset	local	rare
Landslides, avalanches	physical	rapid onset	local	rare
Volcanic eruptions	physical	rapid onset	local/regional	rare

(b) Anthropogenic risk factors	Nature	Timing	Scale	Probability
Pollution Air pollution	chemical	long-term	local/ regional/ global	continuous
Acid rain	chemical	long-term	regional/ global	continuous
Mass tourism	physical/biological/ chemical	long-term	local	continuous

Contd.

Table 4.2 Contd.

(b) Anthropogenic risk factors	Nature	Timing	Scale	Probability
Management				
Heating, air conditioning, ventilation, lighting	physical	long-term	local	continuous
Inappropriate standards of comfort	physical/ chemical	long-term	regional	continuous
Inappropriate use				
Occasional use with sudden warming	physical	long-term	local	continuous
Localised vibration	physical	long-term	local	continuous
Transformation of buildings	architectural	rapid onset	local	frequent
Bad care				
Improper cleaning	physical/ chemical	long-term	local	continuous
Late interventions	physical/ chemical/ biological	medium-term	local	frequent
Careless handling, neglect	physical/ chemical/ biological	medium-term	local	infrequent
Inadequate remedial actions	physical/ chemical/ biological	medium-term	local	infrequent
Remedial works				
Bad choice of experts	physical/ chemical/ biological	rapid onset	local	infrequent
Treatments	physical/ chemical	long-term	local	infrequent
Improper repair	physical/ chemical	long-term	local	infrequent
Inappropriate restoration	physical/ chemical	long-term	local	infrequent
Human hazards				
Economic development	environmental	medium-term	regional/ global	continuous
Governmental policy	no limits	rapid onset/ medium-term	regional/ global	infrequent/ continuous
Accidents, theft, vandalism	physical/ chemical	rapid onset	local	infrequent
Arson, terrorism	physical	rapid onset	local	infrequent
War	physical	rapid onset	regional/ global	rare/ infrequent

wind, snow, sunshine, visibility) are thus characteristic of a particular location, and are predicted or forecast for a specific period of time.

Climatic variables

Britain and much of Europe enjoy a moist temperate climate, typical of
the middle latitudes of the world. Compared to other continents, however,
our climate is variable and subject to extremes. These variations have an
important influence, both on the design of buildings, and on the performance and longevity of the building materials (Fig. 4.5).

- *Rainfall* over England is related to topography, with the wettest areas being
  the Lake District, Pennines and moors of south-west England (average annual rainfall recorded of over 2000 mm), and the driest being East Anglia,
  and parts of the Midlands, north-east and south-east England (average annual rainfall recorded of less than 700 mm). The maximum recorded rainfall
  in any one day was 279 mm at Martinstown (Dorset) on 18 July 1955.
- The mean annual *temperature* at low altitude varies from about 8.5°C to
  11°C, with the highest values occurring around or near to the coast of
  Cornwall; these temperatures decrease by approximately 0.5°C for every
  100 m increase in height. The highest recorded air temperature was 38.5°C
  at Brogdale (Kent) on 10 August 2003, and the lowest was −26.1°C at New-
  port (Shropshire) on 10 January 1982.
- *Wind* speed and direction is affected by topography, roughness of the ter-
  rain, and the presence of buildings and other structures. The highest gust
  recorded at a low-level site was 103 knots (118 mph) at Gwennap Head
  (Cornwall) on 15 December 1979.
- *Sleet or snow* falls on 10 or fewer days each year in some south-western
  coastal areas and on over 50 days in the Pennines; the average number of
  days with snow lying on the ground varies from 5 or less around the coast
  to over 90 in parts of the Pennines.
- Mean daily *sunshine* figures reach a maximum in May or June and are lowest
  in December. Many places along the south coast achieve annual average
  figures of 1750 hours sunshine, while mountainous areas may receive less
  than 1000 hours.
- *Visibility* is generally good, particularly on the coast, moorland and in the
  mountains remote from industrial or populous areas.

Fig. 4.5 Climatic factors (Meteorological Office, 2006).

Local weather conditions are influenced by aspects of the natural and
built environments, with factors such as topography (hills, valleys), vegetation (trees), buildings (situation, orientation) and human activity (pollution) creating and affecting individual microclimates. Each of these may be
used to modify or control the effects of the weather on individual buildings
and on conditions within.

Fig. 4.6   Average rainfall in England during April 1998 was 115.7 mm, making it the wettest of the century. Flooding around the country, such as at Bengeworth in Hereford and Worcester, caused disruption and damage to property. (Picture courtesy of *Evesham Journal Series*, a Newsquest publication.)

More severe weather patterns, including those resulting from climate changes (such as the effects of global warming and El Niño, which disrupts ocean and atmospheric conditions across the Pacific), and other natural phenomena, including avalanches (powder, wet or slab), earthquakes (including after-shocks and associated fires) (see Chapter 6), volcanoes, floods (flash and slow) (Fig. 4.6) (BRE GBG 11, 1997), tidal waves and tornadoes, may also have to be considered in certain parts of the world, and in the future design and construction of buildings.

Lightning

Lightning is formed by a discharge of electricity between two clouds or a cloud and the earth (more than 85% of lightning occurs over land) due to differences in electrical potential. Peak currents average 20 000 amps, and may reach 200 000 amps for a few milliseconds. The lightning flash, made up of both negatively charged leader and positively charged return strokes, heats the surrounding air to 15 000°C. The resulting expansion of the air around the path of the lightning stroke causes the characteristic

Fig. 4.7   Damage caused by lightning at Blickling, Norfolk in June 1993.

sound of thunder, which travels at about 300 m/s and is hence heard after the visible lightning flash, which travels at 2.99×10^8 m/s.

Conduction of the high-voltage lightning discharge to earth is most likely to occur through isolated buildings or elevated conductors, and poses a serious risk where there is resistance to the passage of the strike (Figs 4.7 and 5.3). Damage is caused by the release of heat energy, fire, cracking and melting of metals, and overloading of electrical installations. Protection is given to buildings and structures considered to be of importance or at particular risk through a system of air terminals, low-resistance down conductors and earth terminals (BRE Digest 428, 1998).

Weathering and staining

The process of weathering by sun, wind and rain is defined as the breakdown and alteration of materials by mechanical and chemical processes. Mechanical or physical weathering includes the action of frost and extreme temperature changes, whilst chemical weathering includes the dissolution of materials into solution, carbonation (dissolution by weak carbonic acid formed by the combination of water with atmospheric carbon dioxide), oxidation (chemical combination of atmospheric oxygen), hydration (chemical combination with water) and the breakdown of chemical bonds.

Fig. 4.8   Failure of brickwork caused by the frost action.

Organic weathering, which may involve both chemical and mechanical processes, is caused by plants and animals.

Frost

Since water expands on freezing, any water that has been absorbed into a porous material exerts a pressure on the internal structure of the material when it freezes, causing fracturing and disintegration. This is commonly seen in brick and stone, when the face fails (spalls) and becomes detached (Fig. 4.8). The susceptibility of a material to such damage is related to the size and distribution of its internal structure of capillaries and pores. In order for water to freeze it requires sufficient space in which the ice crystals can nucleate. Materials with a fine pore structure will thus suffer less damage than those with larger pores in which ice crystals can grow.

Light and other electromagnetic radiation

Visible light forms that part of the spectrum of electromagnetic radiation with a wavelength to which the human eye is sensitive (Fig. 4.9). Other parts of the spectrum, such as infrared and ultraviolet, have differing wavelengths and correspondingly different effects and uses. Natural sunlight is made up of 50% visible, 40% infrared and 10% ultraviolet light.

Light and other electromagnetic radiation provide a source of energy that is capable of initiating photochemical reactions, where energy is absorbed and chemical bonds are broken:

- *Ultraviolet* (UV) radiation has sufficient energy to change the electron distribution in materials. In some instances, this energy is high enough to completely remove an electron from the material, causing oxidation or embrittlement. Plastic (PVCu) rainwater goods will, for instance, become brittle and liable to cracking.
- The energy in *visible light* is also sufficient to bring about photochemical change, particularly with certain sensitive objects. Pictures, paintings, fabrics, dyes and stains will typically fade and weaken as they undergo chemical change, depending on the levels of light and length of exposure.
- Low-energy *infrared* (IR) radiation, present in both sunlight and artificial light sources, is insufficient to cause photochemical change. It is, however, a source of radiant heat and responsible for molecular bond vibrations. Such heat energy can cause embrittlement and shrinkage of organic materials as moisture is lost – this can be seen with warped and split window shutters, and distorted veneers.

Monitoring of visible (lux) and ultraviolet light – whether proportional (μW/lumen) or absolute (μW/m^2) – should be undertaken in situations where materials and contents are likely to be susceptible to damage. Hand-held meters and data loggers are available, and recommended levels for various materials should be consulted both during an inspection or survey, and to assess and monitor the efficacy of remedial action.

Air infiltration

Air, together with airborne particles and contaminants, can enter a building either through openings (doors, windows, vents) or by infiltration through cracks and other imperfections in the construction. Whilst the former is usually the result of design decisions to introduce appropriate levels of ventilation for the occupation and use of the building, the latter may reduce

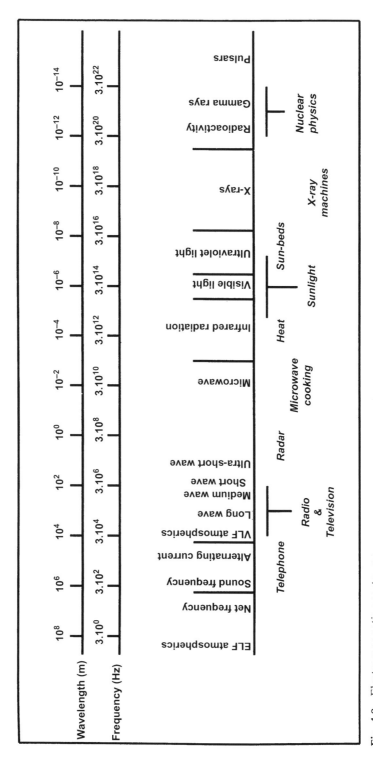

Fig. 4.9   Electromagnetic spectrum.

standards of personal comfort (draughts, introduction of airborne pollutants) and be considered a defect.

Atmospheric gases and pollutants

The air we breathe and that surrounds our buildings contains water vapour,
various pollutant gases and particulate material. These can have an adverse
effect on the individual materials of the building and its contents, and on
the well-being of its occupants.

Although the composition of the air is relatively constant (dry air at
sea level contains 78.08% nitrogen, 20.95% oxygen and other gases), there
may also be various compounds that are hostile to certain materials. The
nature and concentration of these pollutants will depend on such factors
as location, atmospheric and climatic conditions, sources of pollution and
patterns of activity, but may include:

- Sulphur dioxide (SO_2), which forms sulphuric acid in the presence
  of water, will affect calcium carbonate (limestone, marble) (Fig. 4.10),
  cellulose (wood, paper, cotton, linen), proteins (silk, wool, leather,
  parchment), synthetic textiles and materials (nylon, rubber), dyes, pigments and paints. It will also tarnish certain metals.
- Nitrogen oxides (NO_x) can, individually or in combination, have a damaging effect on certain materials. Nitrogen dioxide (NO_2) can act as an
  oxidant and is damaging to cellulose, proteins, and certain pigments
  and dyestuffs.
- Ozone (O_3) is a naturally occurring gas, which protects the earth from
  the damaging effects of cosmic radiation. It is also formed by the action
  of sunlight on exhaust gases (causing photochemical smog) and may be
  given off by certain electrical equipment (such as photocopiers). Ozone
  is a powerful oxidant that is harmful to organic materials – it will cause
  the cracking of rubber, fading of certain dyes and damage to paper.
  On reaction with water it forms hydrogen peroxide, which is a strong
  oxidising agent that will attack cellulose.
- Carbon dioxide (CO_2) dissolved in rainwater, forming carbonic acid,
  can cause the dissociation of calcium carbonate (limestone) to calcium
  bicarbonate.
- Radon (Rn) is a naturally occurring radioactive gas emitted by the successive decay of uranium and thorium within the ground, which further
  decays into radioactive polonium. Both gaseous radon and airborne
  polonium compounds, which can adhere to dust particles (including
  cigarette smoke), can cause severe radiation damage to vital organs following ingestion and increase the risk of lung cancer (Pearce, 1998).

High levels of radon may be found inside buildings (particularly basements and underfloor voids) sited in areas underlain with granite, such
as in Cornwall and Devon, and in parts of Derbyshire, Northamptonshire, Somerset, Wales, Grampian and the Highlands of Scotland (BRE
GBG 25, 1996).

- Gases, such as carbon dioxide (CO_2), methane (CH_4), nitrogen oxides
  (NO_x) and industrial trace gases including chlorofluorocarbons (CFCs),
  hydrofluorocarbons (HFCs) and sulphur hexafluoride (SF_6), are capable of absorbing infrared radiation (greenhouse effect) and contribut-
  ing to an overall warming of the earth and its atmosphere (global
  warming).

Fig. 4.10　Splitting and spalling of limestone surfaces in areas sheltered from direct
rain-washing, due to the formation and subsequent failure of crystalline sulphate
skins.

Particulates present in the air range in size from dusts to tiny particles (particles with a diameter of less than 10 micrometres, known as PM10s, are often used as an indicator of air pollution and are implicated in respiratory diseases), and are typically derived from industrial and vehicular emissions, and the combustion of fossil fuels. Such material may be acidic due to absorbed sulphur dioxide and may contain trace amounts of catalysts that will speed up rates of deterioration. Building materials, such as cement and plaster, release alkaline dust on mixing that can affect certain organic materials. Sea salt (sodium chloride) and other chlorides present in the air can also do damage to porous materials (brick, stone) through crystallisation within the pore structure.

In addition to the common gaseous and particulate pollutants, there are also other airborne contaminants that can affect indoor air quality, and the health and well-being of persons living and working within buildings. Some or all of these may be responsible for individual allergic reactions, such as with respiratory syncytial virus (RSV) and occupational asthma, and be implicated in cases of building-related sickness and sick building syndrome:

- airborne dusts – wood dust, fine sand particles with the potential for causing silicosis, lead carbonate from abrasion of lead-based paints
- mineral fibres – glass fibres, ceramic fibres and asbestos (blue/crocidolite, brown/amosite and white/chrysotile) with the potential for causing mesothelioma, lung fibrosis and cancer
- microbial pollutants – fungi, moulds, algae, pollens, bacteria (such as *Legionella pneumophila*), viruses, protozoa, faecal allergens
- insect pests – house-dust mites (*Dermatophagoides pteronyssinus* and *D. farinae*), psocides (*Liposcelis bostrychophila*)
- micro-organisms and their metabolites – endotoxins, mycotoxins
- allergenic substances – pollen, spores, dust, animal dander, insect faeces, body products
- volatile organic compounds (VOCs) – solvents, chemical treatments (solvents and active ingredients), off-gassing from treated materials, treatment residues (Watt, 1998) (see Chapter 6)
- soot and smoke – open fires, environmental tobacco smoke
- direct soiling – moisture, salts, fats, grease and dirt via inappropriate handling of materials and objects

Research has also shown that illness and stress may be linked to the complex interrelationship between buildings and their occupants, and the artificial environments in which people now live and work (including seasonal affective depression or SAD). A direct link has similarly been shown to exist between productivity and the health and comfort of the workers,

Table 4.3 Chemical, physical and biological factors.

Chemical	Physical	Biological
• oxygen • carbon dioxide • external pollutants • building materials • organic solvents from furniture and furnishings • smoking • substances from cleaning and industrial processes	• temperature • relative humidity • air movement • light (natural and artificial) • electromagnetic radiation • sound and noise • vibration • psychosomatic/psychogenic	• psychology of space and colour • other building occupants • plants and animal pests • microbial factors (fungi, bacteria, viruses, mites, etc.) • workstation ergonomics

including background noise in open-plan office environments (Banbury & Berry, 1998). The importance of chemical, physical and biological factors has been recognised (Singh, 1993, 1996, 1997; Singh & Bennett, 1995) (Table 4.3), and greater regard must be given to such issues in the design and servicing of buildings.

Excess moisture

Excess moisture, caused by rising and penetrating damp, condensation, leakage, spillage or construction processes, is the most widespread and damaging cause of deterioration and decay affecting buildings. The effects of high levels of moisture on the materials from which buildings are constructed can be devastating, as the susceptibility of timber to fungal attack is increased, conditions for chemical and biological degradation are set up, and surface finishes are destroyed (Fig. 4.11) (BRE GBG 5, 1997).

Changes in lifestyle have also created higher levels of moisture. The drive for energy efficiency and personal comfort, with the increased use of insulation products, double glazing and draught-proofing, the move away from open fires, and the desire for higher levels of space heating, all have an effect on the internal climate.

Sources of moisture within a building – such as caused by rising and penetrating damp, condensation, leakage, spillage, building defects and various occupier activities – have to be balanced by moisture sinks through ground drainage, heating, extraction, and natural or artificial ventilation. Where moisture sources are not balanced by appropriate sinks there is the potential for moisture to become trapped within the construction and cause deterioration and eventual decay. Moisture that is not removed by sinks

Fig. 4.11   Vegetation growth due to dampness behind defective hopper head and downpipe. This is of particular concern as the building is not in use and ventilation has been reduced by the use of boarding over the windows.

can be held by porous materials, such as masonry walling, large-section timbers, accumulated rubble, and undrained voids, to form moisture reservoirs. Such reservoirs can retain moisture for long periods, even after the sources have been removed, and contribute to long-term problems of decay.

Rising damp

Water will rise within a wall due to the surface tension (capillarity) between it and the capillaries or pores of porous building materials, and natural osmosis causing water to move from solutions of lower to higher salt concentration (BRE Digest 245, 1989; BRE GBG 6, 1997). Movement of moisture is therefore dependent on the size and distribution of pores within

the wall material, the presence and concentration of soluble salts, and, in theory, the electrical potential of the wall in relation to the surrounding ground.

The presence of excess dampness will increase the risk of fungal attack to skirting boards, floorboards and embedded timbers; cause disruption to internal wall surfaces with staining and damage to finishes; and promote the risk of damage to external walling through salt crystallisation and frost action. Increased levels of associated moisture vapour within a building may also lead to surface or interstitial condensation.

Some of the more common reasons for the presence of rising damp are given in Fig. 4.12.

- Absence or failure of a damp-proof course. It is important to remember that the absence of a damp-proof course does not always mean that the wall(s) will be affected by rising damp, and that such a problem cannot exist where there is a damp-proof course in place.
- Early damp-proof courses of slate may fracture as a result of ground movement, whilst lead or bitumen-based flexible courses may become compressed and so extrude from the joint. Injected courses may be affected by ground-water contaminants or fail to form a continuous band due to the construction of the wall or the rate of moisture movement.
- Damp-proof courses may become bridged externally by an increase in ground levels or pavings; accumulated soil or debris, such as ashes or vegetable matter; or by later deep pointing or render finishes. Internally they may be bridged by inserted solid floors.
- The installation or presence of a damp-proof course in a traditional solid wall construction will theoretically restrict the upward movement of moisture within the wall, but may cause moisture to evaporate out through adjacent solid floors, leading to the premature failure of sealed floor tiles.
- Absence or failure of a link between a damp-proof membrane and damp-proof course. This often occurs where the floor level is lower than the outside ground level, creating a band of unprotected masonry around the base of the wall.
- Presence of hygroscopic salts, often associated with past problems of rising damp, can give misleading indications of current problems unless contaminated plaster is removed or treated on the installation of a damp-proof course.
- Impervious surface finishes (such as cement-based renders and paint systems with high water-vapour resistance) will prevent evaporation of moisture from within a wall, causing it to seek alternative routes for dispersal.
- Hard surfaces around the outside base of a wall will inhibit evaporation of moisture from the ground, and may cause surface water to run back towards the building. Rainwater will also splash up off hard surfaces causing localised saturation of masonry and an increased risk of frost and salt damage.

Fig. 4.12 Common reasons for rising damp.

Penetrating damp

The movement of moisture absorbed into a porous material will depend on the severity of the conditions of exposure, the length of time it is subjected to these conditions and the internal pore structure of the material (BRE GBG 8, 1997). Whether the moisture penetrates the thickness of a wall, and manifests itself on the internal surfaces, will depend on the rate at which it is lost through evaporation to the outside air. Some of the more common reasons for the presence of penetrating damp to consider when inspecting a traditional building are given in Fig. 4.13.

- Damp patches on internal wall surfaces, particularly to exposed south or south-west elevations, where driving rain has saturated the wall or entered the construction through cracks and defective joints in the walling, or defects in external wall finishes. Cavity-wall constructions may be bridged by insulation material, debris or mortar droppings on the ties, or be without damp-proof courses at head, cill and jambs closures.
- Absence or failure of damp-proof trays in cavity walls.
- Failure of render or other surface finishes to exposed faces of parapet walls, upstands and at copings. Saturation of the inside face of a parapet wall in exposed positions may provide a direct route for moisture into the accommodation below.
- Direct penetration of rainwater down an open flue. This will usually show as a damp patch on a chimney breast, often with staining caused by the take-up of soluble soots and tars deposited on the walls of the flue.
- Entry of moisture through open joints and fixing points behind cornices, parapets and balconies.
- Absence or failure of a damp-proof course beneath copings, such as found on parapet walls and gable parapets.
- Absence or failure of flashings and weatherings to features penetrating the roof covering, such as soil and vent pipes, dormer windows, skylights and chimney stacks.
- Damage to, or removal of, projecting features, such as drips, strings or cornices, designed to throw water away from the wall surfaces.
- Where the upper surfaces of projecting features have become eroded, rainwater may collect in such depressions and form localised points of moisture penetration.
- Absence or failure of drips to copings.
- Defective window cills where a drip to the underside is absent or bridged, or the upper surface is angled back towards the building.
- Saturation of a wall may result from inappropriate washing techniques or direct water discharge.

Fig. 4.13 Common reasons for penetrating damp.

Surface and interstitial condensation

The term 'humidity' describes the concentration of water vapour in the atmosphere. This may be defined as either *absolute humidity*, which is the mass of water vapour per unit volume of air (kg/m^3), or *relative humidity*, which is the ratio of moisture in the air to the moisture it would contain if saturated at the same temperature and pressure (% relative humidity or RH).

Relative humidity is inversely proportional to the temperature of the air, so that a drop in temperature will increase the relative level of humidity to a point where the vapour will condense (dew point). The relative humidity measured in the London area in January may thus be 80–90% throughout a 24-hour period, whilst in July this may drop to 60% or less at midday when temperatures are high (Allen, 1997, p. 11).

When air becomes saturated (100% RH), water droplets will condense on cold surfaces, such as glass, tiles, impermeable gloss paint and vinyl wallpaper. This is *surface condensation* (BRE Digest 297, 1985; BRE GBG 7, 1997). Where the dew-point temperature is reached within the thickness of the construction, condensation will form at that point. This is *interstitial condensation* (BRE Digest 369, 1992).

High levels of moisture vapour are generated on a daily basis by cooking (2–4 litres), bathing (0.5–1 litre), clothes drying (3–7.5 litres), and the use of flueless gas and paraffin heaters (1–2 litres during evening), whilst rising damp, rain penetration, leaks and spillages also contribute to increased moisture levels within the building. Natural ventilation may be reduced by the installation of double glazing and draught-proofing, and the blocking of open flues and air bricks.

Some of the more common factors associated with the presence of condensation that may need to be considered when surveying or inspecting a building are given in Fig. 4.14. It is also necessary to bear in mind that standards of human comfort are not the same as those needed for the well-being of sensitive finishes, furnishings and chattels. The comfort range varies according to the relative humidity of the air within the room, being between 20° and 28°C at 30% RH and between 18° and 26.5°C at 70% RH.

The ideal environmental conditions for, say, a mixed material collection in a museum or gallery would, however, be between 16° and 20°C (with a 5° maximum change over 24 hours) and between 50 and 65% RH (with a maximum 5% change over 24 hours). It is therefore necessary to consider how a room or space is used when undertaking an inspection or survey, and to think about what improvements might be

- Mould growth, often found in poorly ventilated areas such as within cupboards or behind furniture, causes damage to finishes, furnishings and clothing, and poses a threat of respiratory problems through the presence of spores.
- High humidities will increase the risks of corrosion and damage to textiles.
- Saturation of thermal insulation will reduce its efficiency and lead to an increase in heating costs.
- Condensation may form on the underside of roofing felt in an inadequately ventilated roof space. The installation or presence of fire stopping, cavity closers and insulation may also restrict ventilation.
- Heating in lightweight modern constructions is typically provided only during the hours of occupation, raising the ambient temperature in a relatively short period of time. If the building is heated only intermittently, the fabric of the building tends to remain cold (thermal inertia). The air, although capable of holding moisture vapour generated by occupation whilst the heating system is on, quickly drops in temperature when the heating is turned off and deposits condensation on cold internal surfaces at dew-point temperature. In traditional constructions, walls and ground floors often have a high mass and therefore will have a slow thermal response (i.e. they warm and cool slowly). Where building fabric has a slow thermal response, constant low-output heating systems are preferred.

Fig. 4.14 Common factors associated with condensation.

made to isolate or protect vulnerable items from exposure to inappropriate conditions.

Chemical, physical and biological action

The materials used in the construction, finishing and furnishing of buildings are subject to the relentless action of chemical, physical and biological factors resulting both from their interaction with the natural environment, and with other materials and substances found in and around the building.

Often such factors will affect particular materials in a manner that is both consistent and predictable, as with the corrosion of ferrous metal or fungal infection of damp timber. In other cases, seemingly innocent changes in construction or building use may trigger unforeseen actions that require detailed and often costly diagnosis. In either case it is essential that the cause or causes of the defect be fully understood before remedial action is specified and implemented.

Many of the chemical, physical or biological actions that affect buildings and materials are, however, complex and require specialised

knowledge in order to correctly identify cause and effect. This section can therefore only draw attention to the main issues and the importance of appropriate specialised advice, and recommend further reading.

Chemical action and change

The chemical elements and compounds that make up materials used in and around buildings are exposed to the constant action of people, processes and the environment. Many of these interactions involve or result in chemical reactions, where materials undergo some form of chemical change resulting in the formation of new compounds. These reactions may be reversible, such that the products can react to give the original reactants, but generally such reversibility is negligible and the reaction is considered to be irreversible.

Examples of the types of chemical action that can adversely affect building materials include the corrosion of metals; sulphate attack on cement-based mortars, render and concrete; and the carbonation of concrete. Awareness of incompatibilities between different materials and the potential for damage to those used for internal finishings and contents are also extremely important and demonstrate the need for vigilance when undertaking inspections and surveys, and when specifying new or replacement materials.

Corrosion of metals

Corrosion occurs due to a reaction between a metal and its environment, and as such can be affected by levels of atmospheric pollutants, concentrations of acids and salts, and the presence or proximity of dissimilar metals and other materials.

Direct oxidation of a ferrous metal is an electrochemical reaction that occurs in the presence of water, oxygen and an electrolyte (electrically conductive liquid), or in air having a relative humidity of over 50% (Fig. 4.15). In urban areas the electrolyte is commonly iron (II) sulphate, which forms as a result of attack by atmospheric sulphur dioxide, whilst in marine areas airborne particles of salt are significant (Greenwood & Earnshaw, 1984, p. 1250).

The corrosion of ferrous metals leads to the formation of hydrated iron (III) oxide, $Fe(OH)_3$ or $FeO(OH)$, giving the well-known red-brown layer of rust (Equation 4.1). This affords a level of protection, but the oxide layer

Fig. 4.15   Corrosion and failure of the connections between the decorative cast-iron rafters and frame members in a mid-nineteenth-century glasshouse.

tends to flake off, so exposing more of the metal to corrosion. Such oxidation also results in a volumetric expansion of the metal – wrought iron is capable of expanding up to seven times its original volume. This expansion may result in movement and damage to other materials (such as that caused by corroding iron cramps in stonework or hoop-iron brickwork reinforcement) (Fig. 4.16).

Cathodic reduction

oxygen + water + electrons → hydroxide ions
$$3O_2 + 6H_2O + 12e^- \rightarrow 12OH^-$$

Anodic oxidations

iron → iron (II) ions + electrons
$$4Fe \rightarrow 4Fe^{2+} + 8e^-$$

iron (II) ions → iron (III) ions + electrons
$$4Fe^{2+} \rightarrow 4Fe^{3+} + 4e^-$$

Fig. 4.16　Corrosion of embedded ironwork resulting in damage to stone.

Overall

iron + oxygen + water → iron (III) ions + hydroxide ions ≡ hydrated iron (III) oxide (rust)

$4Fe + 3O_2 + 6H_2O \rightarrow 4Fe^{3+} + 12OH^- \equiv 4Fe(OH)_3$ or $4FeO(OH) + 4H_2O$

$$(4.1)$$

　　　The electrochemical corrosion of ferrous and non-ferrous metals may also occur when one metal comes into contact with another in the presence of an electrolyte. In such situations the electrochemical reaction takes place where a current passes from one metal to the other – this is how a torch battery produces an electrical current. The metal with the lower potential (anode) will corrode in preference to that with the higher or positive potential (cathode) due to the movement of electrons, and hence electrical charge, from one to the other.

　　　Each metallic element reacts differently to exposure, and to the presence of other metals, according to its position in the electrochemical series (Table 4.4). A metal with a negative potential will thus corrode in the presence of one whose potential is less negative or positive. Examples include the corrosion of zinc-galvanised water cisterns by connected copper piping, and galvanised nails or gang-plates when used in moist timbers treated with copper-based preservatives.

Table 4.4 Standard electrode potentials (E^{θ}) (Volts) with reference to the standard hydrogen electrode.

Element	Electrode potential (V)	Element	Electrode potential (V)
Magnesium ($Mg^{2+} \rightarrow Mg$)	−2.37	Lead ($Pb^{2+} \rightarrow Pb$)	−0.13
Aluminium ($Al^{3+} \rightarrow Al$)	−1.66	Hydrogen ($2H^{+} \rightarrow H_2$)	0.00
Zinc ($Zn^{2+} \rightarrow Zn$)	−0.76	Copper ($Cu^{2+} \rightarrow Cu$)	+0.34
Chromium ($Cr^{2+} \rightarrow Cr$)	−0.71	Mercury ($Hg_2^{2+} \rightarrow 2Hg$)	+0.80
Iron ($Fe^{2+} \rightarrow Fe$)	−0.44	Silver ($Ag^{+} \rightarrow Ag$)	+0.80
Nickel ($Ni^{2+} \rightarrow Ni$)	−0.23	Platinum ($Pt^{2+} \rightarrow Pt$)	+1.20
Tin ($Sn^{2+} \rightarrow Sn$)	−0.14	Gold ($Au^{3+} \rightarrow Au$)	+1.50

This potential difference is also used to give protection to certain metals, where a metal with the lower potential corrodes sacrificially to that with the higher potential. This explains the use of zinc for galvanising steelwork, and sacrificial anodes of zinc, magnesium or aluminium in systems of cathodic protection (see Chapter 6). The exposed surfaces of non-ferrous metals also oxidise, forming a protective layer or patina such as seen on aluminium (aluminium oxide), copper (green copper carbonate or sulphate), lead (blue-grey lead carbonate or sulphate) and zinc (zinc carbonate).

Sulphate attack

Sulphate salts (typically of calcium, magnesium and sodium) are naturally present in the ground, groundwater and various building materials, and also arise from industrial pollution. When such salts come into contact with cement-based mortars, renders and concrete, a reaction occurs between the sulphates and one of the four main components of cement (tricalcium aluminate) resulting in the formation of ettringite (Equation 4.2). This metastable compound takes up large amounts (32 molecules/molecule) of water, which results in expansion, leading to the breakdown and failure of the concrete or cementitious product. This effect can be minimised (such as with sulphate-resisting Portland cement) by reducing the amount of tricalcium aluminate in the cement.

tricalcium aluminate + calcium sulphate + water → ettringite

$$Ca_3Al_2O_6 \quad + 3CaSO_4{\cdot}2H_2O \quad + 26H_2O \rightarrow 3CaO{\cdot}Al_2O_3{\cdot}3CaSO_4{\cdot}32H_2O \quad (4.2)$$

Sulphate attack manifests itself as cracking and spalling of concrete and cement-based renders, and occasionally induces curvature in unlined chimney stacks, where sulphurous flue gases cause differential expansion between mortar joints to the exposed and sheltered sides of the stack.

Carbonation of concrete

A concrete matrix is naturally alkaline due to the presence of calcium hydroxide, and this alkalinity (pH 12.5 to 13.5) confers a level of chemical protection to embedded steel reinforcement. Where acidic atmospheric gases (such as carbon dioxide, sulphur dioxide and sulphur trioxide) are in contact with the concrete and enter the matrix in solution through pores (permeability), cracks and damaged areas, reactions between the alkaline material and acidic solutions result in a reduction in alkalinity known as *carbonation* (Equation 4.3). Carbonated concrete offers little protection to embedded metal, and corrosion may occur. This, in turn, causes cracking and spalling of the concrete, which allows further carbonation to take place.

alcium hydroxide + carbon dioxide → calcium carbonate + water

$$Ca(OH)_2 \quad\quad + CO_2 \quad\quad \rightarrow CaCO_3 \quad\quad + H_2O \quad\quad\quad (4.3)$$

Incompatibilities between different materials

Certain materials used in the construction and finishing of buildings have the potential to react when in contact with one another. Examples of such incompatibility include the corrosive effects of alkali run-off from wet cement, mortar and plaster, or rainwater run-off from copper-clad roofs or from copper pipes, on to aluminium and zinc roof coverings.

Chemical action in internal environments

Chemical reactions may also take place inside buildings, where materials used in finishings, furnishings and chattels are exposed to high levels of atmospheric gases and pollutants. Sulphur dioxide, nitrogen oxides and ozone can, for instance, have serious effects on particular materials.

Internal pollutants, often produced through the day-to-day use of a building, are also potentially damaging – alkaline cement and plaster particles released during building works have been referred to above, but soots and smoke from open fires may also stain porous materials such as marble and stone. Valuable finishes (such as gilded surfaces) and objects should, for this reason, be removed or protected in advance of works likely to cause high levels of dust and pollution.

Chemical action is of particular concern in museums and galleries, where materials used for room finishes, the construction of display cases and mounts, and cleaning can have an adverse effect on artefacts. The artefacts or exhibits themselves may also need to be considered in relation to each other where there could be a reaction between their constituent materials. Examples of such potential reactions include:

- volatile chloride emissions, such as emitted by various plastics and fire-retardant inorganic salts, will affect copper- and iron-based artefacts
- volatile organic (mainly acetic and formic) acids, such as emitted by woods and wood composites (medium-density fibreboard, plywood, blockboard, chipboard), domestic paints, adhesives, varnishes, surface coatings and certain textiles, will affect copper and lead, and attack photographic emulsions
- formaldehyde emitted by certain adhesives (i.e. containing urea or phenol formaldehyde), wood composites, cardboard, foams, paints and certain fabrics will affect metals and organic materials
- hydrogen and carbonyl sulphides, such as emitted by wool, felt, certain fabrics, some rubber adhesives and draught excluders, will tarnish copper and silver

Physical action and change

As well as chemical change, materials may also be affected by heat, moisture, crystallisation of soluble salts, light, sound, electricity and magnetism. These can cause physical changes that do not result in the formation of new compounds, but rather alter the physical character and performance of the affected material for as long as the change persists. The most common physical actions are noted below.

Thermal movement

Materials exposed to varying thermal conditions will undergo dimensional change as a result of the vibration and movement of their atoms and molecules. The extent of this movement is related to the strength and character of the bonding between these atoms or molecules, and to the increase or decrease in temperature. The measurement of movement in a particular material is known as its coefficient of thermal expansion (Table 4.5). Unless restrained, most materials will contract in response to a decrease in temperature.

Where materials are exposed to cycles of high and low temperatures, as experienced during the course of the day (diurnal), between day and night, and between summer and winter (seasonal), the material will experience corresponding cycles of expansion and contraction. Within a homogeneous material (particularly stone), such cyclical movement between the surface and the body of the material can lead to shear failure and the detachment of the surface layer.

Where materials with different coefficients are bonded or secured together, differential expansion and contraction will cause the bond or

Table 4.5 Effects of thermal movement on building materials.

Material	Coefficient of thermal expansion ($\times 10^{-6}$/°C)	Approx. unrestrained movement for a 30°C change in a material length of 3 m (mm)
Timber and plant material		
Oak, along fibres	3.4	0.306
Pine, along fibres	5.4	0.486
Wood laminates	10–40	0.900–3.600
Oak, across fibres	29	2.610
Pine, across fibres	34	3.060
Stone		
Marble	4–6	0.360–0.540
Limestones	4–7	0.360–0.630
Sandstones	5–12	0.450–1.080
Granite	8–11	0.720–0.990
Slate	9–11	0.810–0.990
Ceramics		
Clay bricks	5–8	0.450–0.720
Calcium silicate bricks	12–22	1.080–1.980
Binders and concrete		
Concrete blockwork	6–12	0.540–1.080
Lime mortar	8–10	0.720–0.900
Cement mortar	10–11	0.900–0.990
Gypsum plaster	10–12	0.900–1.080
Concrete	10–14	0.900–1.260
Metals		
Cast iron	10.6	0.954
Steel	10–14	0.900–1.260
Iron	11–12	0.990–1.080
Copper	17	1.540
Stainless steel	10–18	0.900–1.620
Brass	18	1.620
Bronze	20	1.800
Aluminium	24	2.160
Lead	29	2.610
Zinc	31	2.790
Glass		
Glass	7–9	0.630–0.810
Modern materials		
Plastics	14–97	1.260–8.750
Epoxy resins	60	5.400
Acrylic resins	70–80	6.300–7.200
Polyester resins	100–150	9.000–13.500

weaker of the two materials to fail. Similarly, the internal stresses that are generated within a material that is prevented from expanding or contracting due to the use of inappropriate fixings or other forms of restraint will cause it to crack or distort (Fig. 4.17).

Fig. 4.17   Copper sheet used as a covering for the pediment and dormer roof, and lead for the dormer cheeks and as a weathering for the timber cornice. Both sheets metals have suffered from thermal movement on this south-facing section of the roof.

Moisture movement

The absorption of water into porous materials occurs due to a weak chemical/physical bond between the surface of the material and free molecules of water. This absorption typically causes an increase in the volume of the material, as shown in Table 4.6. Conversely, a moisture loss tends to lead to a decrease in volume and corresponding shrinkage. Porous masonry, such as brick and stone, may suffer from both thermal and moisture-related

Table 4.6 Effects of moisture movement on building materials.

Material	Approx. unrestrained movement from dry to saturated for a material length of 3 m (mm)
Wood and plant materials	
Wood, along grain	0.025
Wood, across grain	3–5% radial and 5–15% tangential
Stone	
Marble	Minimal
Ceramics	
Clay bricks	0.075–0.300
Calcium-silicate bricks	0.300–1.500
Binders and concrete	
Gypsum plaster	Minimal
Sand plaster	Minimal
Concrete	0.600–1.800
Metals	
Aluminium	Nil
Copper	Nil
Steel	Nil
Glass	
Glass	Nil
Modern materials	
Plastics	Minimal

movement, and thus be particularly susceptible to deterioration and decay.

Materials that require the addition of moisture during their manufacture, such as mortars, renders and plasters, will contract during setting, and respond little to later wetting. Other materials, such as unseasoned timber, will suffer continual expansion and contraction due to changes in equilibrium moisture content. Moisture absorbed into a porous material may carry with it contaminants that can form chemical compounds leading to varying rates of irreversible expansion. An example of this is the expansive forces that act on concrete and cement-based products under the action of sulphates (see above).

Crystallisation of soluble salts

Where dissolved salts are present within a porous material, evaporation of the solvent will tend to concentrate the salts at the surface, where they will crystallise and form the white deposit known as *efflorescence*. Clay bricks typically contain sulphates of calcium, sodium, potassium, magnesium and iron, whilst contamination from ground water and other sources can also introduce carbonates, nitrates and chlorides. Efflorescence is fairly common on new brickwork as moisture is taken up and released by evaporation. Persistent efflorescence may, however, indicate the presence of excess moisture due to defective detailing or workmanship (Fig. 4.18).

Fig. 4.18   Persistent efflorescence on brickwork where rainwater is not effectively removed from the construction.

Where the salt crystals form within the pores of the material, rather than at the surface, the resulting pressure can set up internal stresses that cause the failure and disintegration of the surface layer. This is sometimes known as *sub-* or *crypto-efflorescence*. Certain salts (such as sodium chloride and sodium nitrate) are also *hygroscopic*, which means that they have the

ability to attract moisture from the atmosphere. This will increase the moisture content of the contaminated material and may be misinterpreted as, say, penetrating damp (see Chapter 6). Other salts (particularly calcium salts) attract moisture to the extent that they will go into solution at high humidities (*deliquescence*), leaving the surface of a wall visibly damp.

Biological action and change

Plants

Trees and other forms of vegetation can affect buildings either by the action of their roots on foundations and underground services (particularly drains) or through direct contact of branches and climbing roots with the walls and roof coverings (BRE Digest 418, 1996; BRE GBG 2, 1997). Climbing plants and creepers, such as ivy (*Hedera helix*), can cause particular damage through the action of their aerial roots, suckers and tendrils, and by the secretion of acidic substances.

Ground conditions may also be affected by the amounts of water taken up by trees and plants. Clay soils are particularly susceptible to fluctuating moisture levels, with shrinkage and swelling causing often substantial ground movement and associated building subsidence. Trees, shrubs and climbing plants may cause blockages in gutters and downpipes, dislodge slates and tiles, shade parts of buildings, and restrict the evaporation of moisture from wall surfaces.

Fungi

Fungal attacks in buildings start with the spores of the fungus landing on timber surfaces. If the timber is damp, these spores germinate and send out hyphal threads (rhizomorphs) to form the mycelium, which feeds on the organic matter in the wood causing it to decay. Within the mycelium a sporophore, or reproductive fruiting body, will develop in time and release further spores which, travelling on air currents, spread the infection to other susceptible timbers (Fig. 4.19).

Fungi that affect timbers in and around buildings are commonly referred to as either 'dry' or 'wet' rots. This terminology is confusing, as all such fungi require the timber to have a moisture content above 20%. Wood-rotting fungi may be further designated as 'brown' or 'white' rots – the former consumes only the pale-coloured cellulose causing the affected wood to darken in colour, whilst the latter consumes both cellulose and lignin, giving a bleached appearance to the affected wood.

In all cases of fungal attack it is important to make a correct identification of the fungus so that appropriate remedial action may be undertaken (BRE

Fig. 4.19   Fruiting body and spores of the dry rot fungus (*Serpula lacrymans*). (Photograph by Mary Manning.)

Digest 345, 1989 and Digest 299, 1993; BRE GBG 12, 1997). It is particularly important to make a clear distinction between the true dry rot fungus, *Serpula lacrymans*, and other wood-rotting fungi, as it is able to spread by sending out strands in search of further timber and cause extensive damage to structural timbers.

The term 'wet rot' is used to describe a number of wood-rotting fungi, the two most common forms of which are cellar fungus (*Coniophora puteana*) and white pore or mine fungus (*Fibroporia vaillantii*) (Fig. 4.20). Other wet rot fungi include *Amyloporia xantha*, *Poria placenta*, *Phellinus contiguus*, *Donkioporia expansa*, *Pleurotus ostreatus*, *Asterostroma* spp., and *Paxillus panuoides*. There are, in addition, various moulds, stains and fungi that should be considered when assessing the condition of timber or masonry, and other moulds and mildews that will attack organic materials such as leather, fabrics and paper. Details of the most common fungi to affect timber in buildings are given in Fig. 4.21.

Lichens, mosses, algae and moulds

Lichens, mosses and other biological growths colonise suitable outside surfaces where they can feed on mineral salts, and as such can provide an attractive appearance to roof coverings and walls. Their presence may,

Fig. 4.20   Mycelium of wet rot (*Fibroporia vaillanti*) (Photograph by Hutton + Rostron).

however, be detrimental to the material on which they are present – acidic metabolic products may damage materials such as limestone, their action on the surface of the material may change the porosity and permeability of the material and so cause micro-cracking and surface erosion, and their growth may obscure carvings and inscriptions (BRE Digest 370, 1992). Mosses may become dislodged and cause blockages in gutters and downpipes. Algae may also appear in wet situations, and these may cause staining to the affected surfaces.

Moulds germinate on damp internal surfaces where there is a source of organic material to sustain their growth. They may appear as stains or distinct spots on affected surfaces, and may be specific to certain processes (such as those performed in bakeries, breweries and maltings). Remedial measures include the use of fungicidal washes and paints, ventilation, dehumidification, insulation and heating.

Birds and vermin

Birds, such as pigeons (*Columba livia*), sparrows (*Passer domesticus*) and starlings (*Sturnus vulgaris*), are capable of causing indirect damage to buildings by dislodging roof coverings, blocking rainwater disposal systems with nesting material, and soiling wall surfaces. They may also pose a nuisance through their calling, nesting activities and scavenging for food. The birds and their nests may also carry fleas, mites and other parasitic insects.

Dry rot (*Serpula lacrymans*)
- Brown rot.
- Attacks mostly softwoods.
- Requires timber with a moisture content of about 25%.
- Needs unventilated space with ambient and surface temperatures of 18–24°C (65–75°F).
- Requires a source of calcium or iron to neutralise the oxalic acid produced through metabolism.
- Deep cross-cracking of timber into cuboidal pieces about 50 mm in size.
- Branching white or grey surface strands up to 5 mm thick, which become brittle when cut and dried.
- Fruiting bodies rusty-red in colour due to spores, with white or grey margin. Texture soft and tough.
- Mycelium forms as soft white cushions or thick grey sheets with yellow patches.

Cellar fungus (*Coniophora puteana*)
- Brown rot.
- Attacks hardwoods and softwoods.
- Requires timber with a moisture content greater than 25%.
- Needs unventilated space with ambient and surface temperatures of 18–24°C (65–75°F).
- Most common form of decay found with saturated timber.
- Cross-cracking with visible darkening of wood. The surface of the timber may appear sound.
- Slender dark-brown or black surface strands.
- Fruiting bodies are rare, but olive-brown in colour, and surface raised in lumps and pimples.
- Mycelium rare except in conditions of high humidity, but form as thin skin-like growths.

White pore or mine fungus (*Fibroporia vaillantii*)
- Brown rot.
- Attacks softwoods.
- Requires timber with a moisture content greater than 25%.
- Needs unventilated space with ambient and surface temperatures of 18–24°C (65–75°F).
- Cuboidal cracking of timber as seen with *Serpula lacrymans*, but less deep.
- White surface strands up to 2.5 mm thick, which remain flexible when cut and dried.
- White or pale-coloured fruiting bodies forming as irregular sheets or plates.
- Mycelium white or cream coloured sheets (Fig. 4.20).

Fig. 4.21 Common fungi to affect timber in buildings.

Feral pigeons are also carriers of diseases such as salmonellosis, psitta-
cosis and pseudo-tuberculosis. Their droppings may increase the rate of
lichen growth and promote the spread of other diseases including histo-
plasmosis, aspergillosis, cryptococcosis and listeriosis (Hutton & Dobson,
1992, pp. 160–61).

Rodents, such as the common or brown rat (*Rattus norvegicus*), black
or ship rat (*Rattus rattus*) and house mouse (*Mus musculus*), cause direct
damage by biting, gnawing and chewing various materials and electrical
cabling, and spread diseases such as leptospirosis (Weil's disease), trichi-
nosis and salmonella poisoning.

Squirrels, typically the grey squirrel (*Sciurus carolinensis*), often build
their nests or dreys in roof spaces or even in chimney flues, and can dislodge roof coverings during their movements and in search for nesting
material.

Bats, although not strictly classed as vermin, can cause indirect damage to materials and finishes within buildings due to the high salt
concentrations in their urine, and oils, fats and fungal spores carried
in their droppings. All 17 species of British bat are protected by the
Wildlife and Countryside Act 1981 as amended, and advice concerning
bats can be obtained from the Bat Conservation Trust or Vincent Wildlife
Trust.

Wood-boring insects and insect pests

The life cycle of wood-boring insects found in buildings starts with eggs
laid on or near the surface of timber. The resulting larvae tunnel into the
timber to feed on the available organic material and, depending on the
species and the prevailing conditions, may spend between one and eleven
years there. Eventually, the larva forms a pupal chamber close to the surface
of the timber, from which the adult beetle emerges causing the characteristic
flight hole and bore dust (BRE Digest 307, 1992; BRE GBG 13, 1998).

Other insects that can cause damage to structural and non-structural
timbers used within a building include weevils, wasps and borers. Termites are not usually present in buildings located in cool temperate climates, although recent infestations of the European subterranean termite
(*Reticulitermes lucifugus*) in Devon have been documented (Eggleton, 1994;
Coggins, 1997; Hollis, 1998; Eggleton *et al.*, 1998; BRE Digest 443, 1999). All
these insects may also be inadvertently introduced within timber objects
and plants from abroad.

Details of the most common wood-boring insects to affect buildings and
their contents are given in Fig. 4.22. Insect pests may also cause damage
to furnishings and chattels, whether in a museum, gallery or in the home
(Pinniger, 1994). These include carpet beetles (and their larvae known as

Common furniture beetle *(Anobium punctatum)*
- Commonly referred to as 'woodworm'.
- Attacks softwoods and European hardwoods.
- Requires timber with a moisture content in excess of 12%.
- Flight holes circular, 1–2 mm in diameter.
- Adult beetle 3–5 mm in length.
- Flight season April to August.
- Three-year life cycle.
- Bore dust fine, white and granular or cylindrical pellets pointed at one or both ends.

Deathwatch beetle *(Xestobium rufovillosum)*
- Attacks decaying hardwoods, mostly oak, and adjacent softwoods (Fig. 4.23).
- Flight holes circular and 3 mm diameter.
- Flight season April to June.
- Adult beetle 6–9 mm long.
- Five to twelve-year life cycle.
- Bore dust coarse, bun-shaped pellets.

House longhorn beetle *(Hylotrupes bajulus)*
- Attacks sapwood of most softwoods.
- At present found only in the south-east of England, where special provisions within the Building Regulations apply. Records of infestations in the United Kingdom are maintained by the Building Research Establishment (Lea, 1994).
- Flight holes oval, often ragged, 6–10 mm in diameter.
- Flight season July to September.
- Adult beetle 10–20 mm long.
- Three to eleven-year life cycle.
- Bore dust large, cylindrical sausage-shaped pellets.

Powderpost beetle *(Lyctus brunneus and L. linearis)*
- Attacks sapwood of new hardwoods, particularly oak and elm.
- Flight holes circular 1–2 mm diameter.
- Flight season June to August.
- Adult beetle 4–7 mm long.
- Ten-month life cycle.
- Bore dust soft and silky, very fine powder.

Fig. 4.22 Common wood-boring insects to affect buildings and contents.

Contd.

Ptilinus beetle (*Ptilinus pectinicornis*)
- Attacks certain European hardwoods, particularly beech, elm, hornbeam and maple, and found mainly in furniture.
- Flight holes circular 1–2 mm in diameter.
- Flight season May to July.
- Adult beetle 4-6 mm long.
- Two-year life cycle.
- Bore dust fine powder.

Wood-boring weevil (*Pentarthrum huttoni and Euophryum confine*)
- Attacks decaying hardwoods and softwoods.
- Flight holes small and ragged 1 mm in diameter.
- Flight season any time.
- Adult weevil 3–5 mm long.
- Seven to nine-month life cycle.
- Bore dust fine round cylindrical pellets.

Fig. 4.22 Contd.

Fig. 4.23   Damage caused by deathwatch beetle (*Xestobium rufovillosum*) to roof timber.

'woolly bears'), hide and leather beetles, clothes moths, biscuit beetles, book lice, silverfish and mites. Silverfish, wood lice and slugs are typically associated with damp conditions, and their presence inside a building may indicate a potential problem.

Insects such as flour beetles, cockroaches, ants, wasps and flies may pose a nuisance where food is being stored, prepared or consumed, and should be dealt with accordingly. Fleas, bed-bugs and house-dust mites can make life unpleasant for building occupants, but spiders, often the cause of consternation, may provide a welcome service in controlling numbers of other insect pests.

Outside the building, masonry or mortar bees (*Osmia rufa*) may cause damage to soft bricks, stone, mortar joints and clay-based walling (BRE TIL 64, 1982; BRE Digest 418, 1996). The female bees bore into the material during the spring to form a system of galleries or tunnels, and the new brood emerge in early summer. Insecticidal sprays or injection treatments provide only limited protection, and consideration may need to be given to regular spray treatments in early spring or the use of a protection render finish to the wall.

For buildings situated close to areas of vegetation where mammals such as mice and deer are present, infected ticks may spread the bacterium *Borrelia burgdorferi* that can cause Lyme disease in humans.

Movement

The movement of buildings take place both at a molecular level, caused by the response of materials to stimuli such as temperature and moisture (see above), and at a general level due either to the load imposed by the building on the ground (settlement) or some external factor that affects the loadbearing qualities of the ground (subsidence). The correct identification of the underlying causes of building movement is essential if appropriate remedial action is to be undertaken.

Ground conditions

The interaction between a building and its site, and particularly the ground on which it stands, is of utmost importance owing to the variability of ground conditions and various factors that might cause movement and instability. Whilst the foundations of new buildings are designed specifically to take account of such issues, those of existing buildings are often inadequate by present standards and susceptible to changes brought about by increased loadings and changes in site usage.

It is therefore essential that consideration be given to the site and ground conditions as part of an inspection or survey, combining both background research and site investigation. Factors to be taken into account include:

- geological faults (e.g. earthquake zones)
- soil types (e.g. alluvial, shrinkable clay, sandy, silty)
- sloping ground
- land instability
- areas of backfill
- vegetation (e.g. trees, shrubs)
- ditches and watercourses (i.e. past and present)
- ground water (e.g. level of water-table, presence of underground streams)
- mines and natural cavities (e.g. swallow holes caused by localised erosion of chalk or limestone)
- former or existing underground workings
- structures below ground
- contaminated ground (e.g. sulphates)
- risk of vibration from traffic and machinery

Settlement

Settlement is 'movement within a structure due to the distribution or redistribution of loading and stresses within the various elements of construction' (Institute of Structural Engineers, 1994, p. 10). This normally occurs in the early stages in the life of a building and may be associated with the compaction of, or movement in, the ground beneath the foundations due to the self-weight of the building.

Differential settlement may occur where a structure is founded in varying ground conditions or over 'hard shoulders', such as parts of basements, or has foundations imposing unequal loads as might occur with extensions, projecting bays or porches.

Subsidence

Subsidence is the 'downward movement of a building foundation caused by the loss of support of the site beneath the foundations' (Institute of Structural Engineers, 1994, p. 10). This is usually associated with volumetric changes in the subsoil (compression of peat, reactivation of fill, shrinkage in clay soils), brought about as a direct result of some external factor (mining, extraction, change in ground-water levels due to abstraction or land drainage, erosion of fine particles because of leaking drains or underground streams, influence of trees on shrinkable soils).

As with settlement, differential subsidence may occur as a result of varying ground conditions.

Heave

Heave is the 'volumetric expansion of the ground beneath part or all of the building . . . normally taken to mean the upwards movement of the site caused by the expansion or swelling of the subsoil' (Institute of Structural Engineers, 1994, p. 11). Such movement may result from the removal of load, frost or moisture increase in dried clay soils. The freezing of ground water and consequent formation of ice lenses in frost-susceptible soils can cause damage as a result of both the initial heave of the ground and later settlement as the ice thaws.

Landslip

Landslip is the 'sudden movement of soil on a slope, or gradual creep of a slope over a period of time' (Institute of Structural Engineers, 1994, p. 11). Clay hillside soils will, for instance, often exhibit a downhill creep associated with seasonal changes in moisture content.

Foundation failures

Building movement may also occur as a result of the failure of its foundations. This may be caused by changes in ground conditions affecting early footings, failure of foundation arches, decay of timber piles or chemical attack on concrete foundations. Such failures may develop over long periods of time and be associated with other causes of movement (BRE GBG 1, 1997).

Fire

Fire is an oxidation reaction requiring fuel, oxygen and a source of ignition, which is accompanied by heat, light and sound. A fire starts with either pilot or spontaneous ignition of gases given off by a heated material when these are at ignition temperature. The flames that also accompany a fire are formed by burning gases or vapours given off by the material; where no vapours are produced the material may simply burn, but without flames.

The risk of fire has been a particularly significant factor in the design and construction of buildings throughout history, both with the use of non-combustible materials and with the introduction of fire-resistant forms of construction. The requirements of today's regulations for improved protection, fire fighting and means of escape, and the approaches taken by designers to reduce the incidence and effects of fire through improved

Fig. 4.24 Damage caused by fire within the Cartoon Gallery of Hampton Court Palace in April 1986. (Crown copyright: Historic Royal Palaces.)

methods of detection and management (people and resources), are in essence a continuation of this uneasy relationship between humans and fire (Fig. 4.24).

Fire damage

The damage caused by a fire in a building results from both the direct burning of combustible materials and the indirect effects of heat and smoke. There is typically further damage caused by the use of water to extinguish a fire, including the formation of moisture reservoirs, and the serious risk of future fungal decay following refurbishment (Singh, 1991) (Fig. 4.25).

Fig. 4.25   Plaster fungus (*Peziza* sp.) in damp conditions following fire.

Building materials will vary in their response to fire depending on their composition, the manner in which they have been prepared or manufactured, how and where they have been put to use in the building, and what finishes or treatments have been applied. Concrete, for instance, is virtually incombustible (depending on the nature of the aggregate used), whereas steel requires protection (often provided by concrete) to avoid it distorting and buckling when heated under load. Furnishings and contents within a building are typically highly combustible and provide the fuel needed in the initial stages of a fire.

The effects of fire on individual elements or components of a building, and on the building as a whole, may therefore be summarised as follows:

- loss of strength through material decay
- loss of stability caused by excess loading
- failure of framing members or joints and loss of rigidity
- changes to load paths and loss of equilibrium
- inadequate bracing and stiffness
- inadequate robustness for purpose
- expansion and subsequent deformation and cracking

Human factors

The activities and behaviour of humans, whether at the time of design and construction or later during occupation and use, will have a significant effect on the well-being of buildings. Throughout its life, a building will often be exposed to harsh and often aggressive conditions – whether through direct action or neglect – that will erode its materials, wear out its services, and reduce its usefulness for future owners/users.

Attitudes to repair and maintenance, particularly of owner occupiers, are often detrimental to the well-being of a property. Work is typically done on an *ad hoc* basis, reacting to damage and decay rather than proactively planning work in anticipation of failure (Sadler & Ward, 1992). Owners, occupiers and visitors move in and around buildings, subjecting them to patterns of often relentless wear, whilst those charged with looking after the building (cleaning, maintenance, repair) bring about further, although well-meaning, erosion and change.

The uses to which buildings are put may also impose substantial pressures by way of excessive loadings, vibration, harsh environmental conditions, and physical changes that compromise the structural or material basis on which the building was designed and constructed. There are, in addition, the actions of those determined to cause deliberate damage or offence through arson, vandalism, graffiti and terrorism that need to be considered (Fig. 4.26). The risk of such unpredictable action makes it all the more difficult to avoid and/or control.

It may thus be seen that human factors need to be taken into account both during an inspection or survey and, later, when proposing remedial action. In the case of such action, additional factors may need to be considered:

- Inherently poor design and material specification
    - failure to understand and appreciate building materials and their limitations
    - failure to understand and appreciate methods of construction
    - failure to understand basic structural principles
    - failure to understand purpose of design decisions
    - failure to allow for maintenance acccess
    - use of poor-quality materials
- Poor workmanship and supervision
    - poor-quality construction standards
    - inappropriate site practices
    - inadequate supervision
    - inadequate preparation

Fig. 4.26   Terrorist bomb damage to the Church of St Ethelburga in Bishopsgate, London, in April 1993. (Photograph by Society for the Protection of Ancient Buildings.)

- ○ inadequate protection against sun and rain
- ○ inadequate site storage facilities
- ○ inappropriate material storage and handling
- ○ failure in communications between adviser, contractor and sub-contractors
- • Inappropriate alterations and adaptations
  - ○ misuse and overloading
  - ○ unsympathetic changes of use
  - ○ inappropriate interventions
  - ○ general and specific wear and tear
  - ○ impact and vibration
  - ○ obsolescence (i.e. functional, legal, physical, social)
  - ○ old age and redundancy
- • Inadequate use and aftercare
  - ○ failure to carry out routine maintenance and repairs
  - ○ inappropriate design and material specification for maintenance and repairs
  - ○ unexpected user activities
  - ○ inappropriate cleaning routines
  - ○ environmental degradation (litter, graffiti, vandal damage, excrement)

References

Allen, W. (1997) *Envelope Design for Buildings*. Oxford: Architectural Press.

Banbury, S. & Berry, D. (1998) Disruption of office-related tasks by speech and office noise. *British Journal of Psychology*, **89** (3), 499–517.

Bonshor, R. & Harrison, H. (1982) *Quality in Traditional Housing – Volume 1: An Investigation into Faults and their Avoidance*. London: BRE/HMSO.

Building Research Establishment (1982) *Damage Caused by Masonry or Mortar Bees*. Technical Information 64. Garston: BRE.

Building Research Establishment (1985) *Surface Condensation and Mould Growth in Traditionally-Built Dwellings*. Digest 297. Garston: BRE.

Building Research Establishment (1988) *Common Defects in Low-Rise Traditional Housing*. Digest 268. Garston: BRE.

Building Research Establishment (1989) *Rising Damp in Walls: Diagnosis and Treatment*. Digest 245. Garston: BRE.

Building Research Establishment (1989) *Wet Rots: Recognition and Control*. Digest 345. Princes Risborough: BRE.

Building Research Establishment (1992) *Identifying Damage by Wood-Boring Insects*. Digest 307. Princes Risborough: BRE.

Building Research Establishment (1992) *Interstitial Condensation and Fabric Degradation*. Digest 369. Garston: BRE.

Building Research Establishment (1992) *Control of Lichens, Moulds and Similar Growths*. Digest 370. Princes Risborough: BRE.

Building Research Establishment (1993) *Dry Rot: Its Recognition and Control*. Digest 299. Princes Risborough: BRE.

Building Research Establishment (1996) *Bird, Bee and Plant Damage to Buildings*. Digest 418. Garston: BRE.

Building Research Establishment (1996) *Buildings and Radon*. Good Building Guide 25. Garston: BRE.

Building Research Establishment (1997) *Cracks Caused by Foundation Movement*. Good Building Guide 1. Garston: BRE.

Building Research Establishment (1997) *Damage to Buildings Caused by Trees*. Good Building Guide 2. Garston: BRE.

Building Research Establishment (1997) *Diagnosing the Causes of Dampness*. Good Building Guide 5. Garston: BRE.

Building Research Establishment (1997) *Treating Rising Damp in Houses*. Good Building Guide 6. Garston: BRE.

Building Research Establishment (1997) *Treating Condensation in Houses*. Good Building Guide 7. Garston: BRE.

Building Research Establishment (1997) *Treating Rain Penetration in Houses*. Good Building Guide 8. Garston: BRE.

Building Research Establishment (1997) *Repairing Flood Damage – Part 1: Immediate Action; Part 2: Ground Floors and Basements; Part 3: Foundations and Walls; Part 4: Services, Secondary Elements, Finishes, Fittings*. Good Building Guide 11. Garston: BRE.

Building Research Establishment (1997) *Wood Rot: Assessing and Treating Decay*. Good Building Guide 12. Garston: BRE.

Building Research Establishment (1998) *Wood-Boring Insect Attack – Part 1: Identifying and Assessing Damage; Part 2: Treating Damage.* Good Building Guide 13. Garston: BRE.

Building Research Establishment (1998) *Protecting Buildings Against Lightning.* Digest 428. Garston: BRE.

Building Research Establishment (1999) *Termites and UK Buildings – Part 1: Biology, Detection and Diagnosis; Part 2: Control and Management of Subterranean Termites.* Digest 443. Garston: BRE.

Camuffo, D. (1997) Perspectives on risks to architectural conservation. In *Saving Our Architectural Heritage: The Conservation of Historic Stone Structures* (N.S. Baer and R. Snethlage, R. eds), pp. 63–92. Chichester: John Wiley & Sons.

Coggins, C. (1997) Now global warming helps termites to cross Channel, *Building Engineer*, **72** (11), December, 6–7.

Construction Quality Forum (1997) *CQF Database Analysis – Report 4: April 1997.* Garston: CQF.

Eggleton, B. (1994) Termites in North Devon!. *Professional Treater*, **4**, 6.

Eggleton, B., Coggins, C., Brooks, F. & Lloyd, J. (1998) Termites. *Professional Treater*, **18**, 6–7.

Greenwood, N. & Earnshaw, A. (1984) *Chemistry of the Elements.* Oxford: Pergamon Press.

Hollis, M. (1998) Sex and the single termite. *The Valuer*, **67** (4), 24.

Housing Association Property Mutual (1997) *Feedback from Data 1991–1994.* Technical Note No 7. London: HAPM.

Hutton, T. & Dobson, J. (1992) The control of feral pigeons: An independent approach. *Structural Survey*, **11** (2), Autumn, 159–167.

Institute of Structural Engineers (1994) *Subsidence of Low Rise Buildings.* London: SETO Limited.

International Organization for Standardization (1984) *Performance Standards in Buildings: Principles for the Preparation and Factors to be Considered.* ISO 6241. Geneva: ISO.

Lea, R (1994) *House Longhorn Beetle: Geographic Distribution and Status in the UK.* Information Paper 8/94. Garston: BRE Press.

Meteorological Office (2006) The climate of England, www.met-office.gov.uk/climate/uk/location/england/index.html, accessed February 2006.

Pearce, F. (1998) Undermining our Lives?, *New Scientist*, **157** (2125), 20–21

Pinniger, D. (1994) *Insect Pests in Museums*, 3rd edn. London: Archetype Publications.

Sadler, R. & Ward, K. (1992) *Owner Occupiers' Attitudes to House repairs and Maintenance.* Research study undertaken for the Building Conservation Trust. London: Upkeep.

Singh, J. (1991) Preventing decay after the fire. *Fire Prevention*, **244**, November, 26–29.

Singh, J. (1993) Biological contaminants in the built environment and their health implications. *Building Research and Information*, **21** (4), 216–24.

Singh, J. (1996) Health, comfort and productivity in the indoor environment. *Indoor Built Environment*, **5**, 22–33.

Singh, J. (1997) *Historic Building Pathology and Health*. Paper presented at The Health of our Heritage conference, 2nd RIBA National Conservation Conference, 9 May, Bath.

Singh, J. & Bennett, C. (1995) Microbial contaminants in the indoor environment and their health implications. Paper 22, CIBSE Annual Conference.

Trotman, P. (1994) *An Examination of the BRE Advisory Service Database Compiled from Property Inspections*. Paper presented at the Dealing with Defects in Buildings symposium, CIB/ICITE-CNR/DISET, 27–30 September, Varenna, Italy.

Watt, D.S. (1998) Risks associated with chemical treatment residues in buildings. *Structural Survey*, **16** (3), 110–19.

Watt, D. & Colston, B. (2003) Research module 6: Maintenance education and training for listed buildings. www.maintainourheritage.co.uk/pdf/module-6main.pdf, accessed October 2006.

Further reading

Addleson, L. (1972) *Materials for Buildings – Volume 1: Physical and Chemical Aspects of Matter and Strength of Materials; Volume 2 – Water and its Effects (1); Volume 3: Water and its Effects (2)*. London: Iliffe Books.

Addleson, L. (1992) *Building Failures: A Guide to Diagnosis, Remedy and Prevention*. 3rd edn. Oxford: Butterworth Architectural.

Addleson, L. & Rice, C. (1991) *Performance of Materials in Buildings: A Study of the Principles and Agencies of Change*. Oxford: Butterworth-Heinemann.

Allen, W. (1997) *Envelope Design for Buildings*. Oxford: Architectural Press.

Anderson J. & Howard, N. (2000) *The Green Guide to Housing Specification*. Garston: BRE Press.

Andrew, C. (1994) *Stone Cleaning: A Guide for Practitioners*. Edinburgh: Historic Scotland and Robert Gordon University.

Ashurst, J. & Dimes, F. (1998) *Conservation of Building and Decorative Stones*. 2nd edn. London: Butterworth-Heinemann.

Ashurst, N. (1994) *Cleaning Historic Buildings – Volume 1: Substrates, Soiling and Investigation; Volume 2: Cleaning Materials and Processes*. London: Donhead Publishing Ltd.

Atkinson, M. (2000) *Structural Defects Reference Manual for Low-Rise Buildings*. London: Taylor & Francis.

Bravery, A.F., Berry, R.W., Carey, J.K. & Cooper, D.E. (2003) *Recognising Wood Rot and Insect Damage in Buildings*, 3rd edn. Garston: BRE.

Berry, R.W. (1994) *Remedial Treatment of Wood Rot and Insect Attack in Buildings*. Report 256. Garston: BRE.

Brownhill, D. & Rao, S. (2002) *A Sustainability Checklist for Developments: A Common Framework for Developers and Local Authorities*. Garston: BRE Press.

Building Research Establishment (1990) *Decay and Conservation of Stone Masonry*. Digest 177. Princes Risborough: BRE.

Building Research Establishment (1991) *Why do Buildings Crack?* Digest 361. Garston: BRE.

Building Research Establishment (1991) *Structural Appraisal of Existing Buildings for Change of Use*. Digest 366. Garston: BRE.

Building Research Establishment (1996) *Reducing the Risk of Pest Infestations in Buildings*. Digest 415. Garston: BRE.

Building Research Establishment (1997) *Repairing Timber Windows: Investigating Defects and Dealing with Water Leakage*. Good Repair Guide 10. Garston: BRE.

Building Research Establishment (1998) *Re-Covering Pitched Roofs*. Good Repair Guide 14. Garston: BRE.

Building Research Establishment (1998) *Repairing Chimneys and Parapets*. Good Repair Guide 15. Garston: BRE.

Burkinshaw, R. & Parrett, M. (2003) *Diagnosing Damp*. Coventry: RICS Books.

Cassar, M. (1995) *Environmental Management*. London: Routledge.

Construction Audit Limited (1999) *HAPM Workmanship Checklists*. London: Taylor & Francis.

Cook, G.K. & Hinks, A.J. (1992) *Appraising Building Defects: Perspectives on Stability and Hygrothermal Performance*. London: Longman Scientific & Technical.

Cook, G. & Hinks, J. (1997) *The Technology of Building Defects*. London: Taylor & Francis.

Cooke, R.U. & Gibbs, G.B. (1993) *Crumbling Heritage? Studies of Stone Weathering in Polluted Atmospheres*. Wetherby: National Power plc and PowerGen plc.

Curwell, S.R. & March, C.G. (eds) (1986) *Hazardous Building Materials: A Guide to the Selection of Alternatives*. London: E. & F.N. Spon.

Curwell, S.R., March, C.G. & Venables, R. (eds) (1990) *Buildings and Health: The Rosehaugh Guide to the Design, Construction, Use and Management of Buildings*. London: RIBA Publications.

Department of the Environment (1993) *Householders Guide to Radon*. 3rd edn. London: HMSO.

Douglas, J. & Singh, J. (1995) Investigating dry rot in buildings. *Building Research and Information*, **23** (6), 345–52.

Driscoll, R. (1995) *Assessment of Damage in Low-Rise Buildings with Particular Reference to Progressive Foundation Movement*. Garston: BRE.

Eldridge, H.J. (1976) *Common Defects in Buildings*. London: HMSO.

Douglas, J. & Stirling, J.S. (1997) *Dampness in Buildings*. 2nd edn. Oxford: Blackwell Science.

Feilden, B. (1994) *Conservation of Historic Buildings*. 2nd edn. London: Butterworth-Heinemann.

Feld, J. & Carper, K. (1997) *Construction Failure*. London: John Wiley & Sons.

Fire Protection Association (1992) *Fire Protection in Old Buildings and Historic Town Centres*. London: Fire Protection Association.

Fire Protection Association (1995) *Heritage Under Fire: A Guide to the Protection of Historic Buildings*. 2nd edn. London: Fire Protection Association.

Garvin, S.L., Phillipson, M.C. & Sanders, C.H. (1998) *Impact of Climate Change on Buildings*. Report BR 349. Garston: BRE.

HAPM Publications Limited (2001) *Guide to Defect Avoidance*. London: Taylor & Francis.

Hendry, A. & Khalef, F. (2000) *Masonry Wall Construction*. London: Taylor & Francis.

Hutton, T., Lloyd, H. & Singh, J. (1991) The environmental control of timber decay. *Structural Survey*, **10** (1), 5–20.

Levy, M. & Salvadori, M. (1992) *Why Buildings Fall Down: How Structures Fail*. New York: W.W. Norton.

London Hazards Centre (1990) *Sick Building Syndrome: Causes, Effects and Control.* London: London Hazards Centre.

McMullan, R. (1998) *Environmental Science in Building.* 4th edn. Basingstoke: Macmillan Press.

Oliver, A. (1997) *Dampness in Buildings*, 2nd edn. Oxford: Blackwell Science.

Pain, S. (1997) Knock Knock, Who's There? *New Scientist*, **156** (2113/2114), 20–27 December, 52–54.

Phillipsen, M. (2002) The problem of moisture in buildings. *Constructing the Future*, **13**, 5–6.

Property Services Agency (1989) *Defects in Buildings.* London: HMSO.

Ransom, W. (1987) *Building Failures: Diagnosis and Avoidance*, 2nd edn. London: Taylor & Francis.

Richardson, B.A. (2000) *Defects and Deterioration in Buildings: A Practical Guide to the Science and Technology of Material Failure.* London: Taylor & Francis.

Richardson, B.A. (1995) *Remedial Treatment of Buildings.* Oxford: Butterworth-Heinemann.

Ridout, B. (1992) *An Introduction to Timber Decay and its Treatment.* Romsey: Scientific & Educational Services.

Ross, K. (2005) *Modern Methods of Construction: A Surveyor's Guide.* Garston: BRE Press.

Schaffer, R. (2004) *The Weathering of Normal Building Stones* (facsimile of 1972/1932 edition) Shaftesbury: Donhead Publishing.

Singh, J. (1993) Biological contaminants in the built environment and their health implications. *Building Research and Information*, **21** (4), 216–224.

Singh, J. (ed.) (1994) *Building Mycology: Management of Decay and Health in Buildings.* London: E. & F.N. Spon.

Singh, J. & Walker, B. (eds) (1996) *Allergy Problems in Buildings.* Dinton: Quay Books.

Simpson, J. & Horrobin, P. (eds) (1970) *The Weathering and Performance of Building Materials.* Aylesbury: Medical and Technical Publishing.

Smith, B.J. (1976) *Construction Science – Volumes 1 and 2.* London: Longman.

Smith, B. & Turkington, A. (eds) (2004) *Sone Decay: Its Causes and Controls.* Shaftesbury: Donhead Publishing.

Smith, B. & Warke, P. (eds) (1996) *Process of Urban Stone Decay.* Shaftesbury: Donhead Publishing.

Thomson, G. (1986) *The Museum Environment.* London: Butterworth.

Torraca, G. (1988) *Porous Building Materials: Materials Science for Architectural Conservation.* 3rd edn. Rome: ICCROM.

Trotman, P., Sanders, C. & Harrison, H. (2004) *Understanding Dampness: Effects, Causes, Diagnosis and Remedies.* Watford: BRE Bookshop.

Watt, D.S. & Swallow, P.G. (1996) *Surveying Historic Buildings.* Shaftesbury: Donhead Publishing.

Weir, H. (1997) *How to Rescue a Ruin – By Setting Up a Local Buildings Preservation Trust.* 2nd edn. London: Architectural Heritage Fund.

Webster, R.G.M. (ed.) (1992) *Stone Cleaning and the Nature, Soiling and Decay Mechanisms of Stone.* London: Donhead Publishing.

Chapter 5
Survey and Assessment

Fault-finding

The survey and assessment of buildings, which combine the diagnosis of defects in the structure, fabric and services, together with the forecasting of how such problems might develop in the future, requires a detailed understanding of the building. In particular, the surveyor should be aware of how the building was constructed, how is has been used, what alterations have been undertaken, and how it has been repaired and maintained.

Defects are discovered either by the occupants through the manifestation of an obvious fault (such as a leaking roof) or during an intentional inspection or survey of the building by a building professional. It is this logical and systematic approach to fault-finding through survey, diagnosis and prognosis that is considered in this chapter.

Building inspections and surveys

The principal means of obtaining information about the construction and condition of a building comes from undertaking an inspection or survey. For the purposes of this book, a building survey may be defined as 'a comprehensive, critical, detailed and formal inspection of a building to determine its condition and value, often resulting in the production of a report incorporating the results of such an inspection' (Dickinson, 1996, 1/1). The form and content of such a survey usually follows a prescribed format, such as that used for the quinquennial inspection of churches, or is agreed in advance by the client and surveyor.

A building survey usually takes the form of a preliminary site visit, background research, detailed on-site survey, and preparation of a written report. The nature and extent of these activities will vary, depending on the requirements of the client and the form of the building or structure being surveyed.

Preliminary site visit

The purpose of a preliminary site visit or reconnaissance is to inform the surveyor about the nature and likely extent of the survey and what additional considerations need to be taken into account when confirming and undertaking the survey. Such a visit will allow the following:

- make contact with occupants, neighbours and other persons associated with the building
- familiarisation of layout (e.g. dead areas, ducts, flues)
- confirm security arrangements (e.g. disarming and re-arming alarm systems, keys to doors and windows)
- ensure facilities are available for safe access to all parts of the building
- establish nature and extent of services (e.g. availability of electricity and water)
- agree nature and extent of opening up and moving of contents
- establish boundaries, easements and rights of way
- assess need for specialised survey services (e.g. drain and electrical testing; non-destructive survey)
- assess need for specialised investigative equipment (e.g. hoists) (Fig. 5.1)
- take photographs and sketches as a record for later survey

Background research

Research undertaken as part of a survey should be appropriate for its purpose and aim to inform the surveyor about issues concerning location, site, construction, use and occupation of the building. Key issues might include:

- determining whether the building and/or site is legally protected (e.g. listed, scheduled or in a conservation area)
- establishing former uses (e.g. potential hazards)
- establishing site conditions (e.g. flooding, clay subsoils)
- establishing earlier programmes of repair and/or maintenance (see Chapter 6)
- ascertaining development policies with local authorities (e.g. structure and local plans)

Such information will come typically from various primary and secondary sources (Fig. 5.2), and will require interpretation and assessment as to its relevance to the building and the survey being undertaken.

Fig. 5.1 Mobile hoist used in the preliminary inspection of an unstable masonry structure.

Inspection or survey

The inspection or survey of a building is a complex task made up of a number of discrete and interwoven activities that provide information on which to make an assessment of its condition and fitness for purpose as prescribed by relevant documents or individual need. Such a survey will typically include the following (Fig. 5.4).

A building survey will draw on a surveyor's skills of observation and judgement, knowledge derived from training and continuing professional development, practical experience, familiarity with the type of property or particular defect, and an inquiring mind. The surveyor should be competent to perform the required service(s) and clear as to the precise purpose and extent of the survey. Explicit instruction, briefing and documentation are, in this respect, essential.

Personal safety and comfort is also an important consideration, and the hazards that can arise when inspecting or surveying a building are given in Appendix B.

- **Background and historical information**
  - ○ national and local repositories (e.g. National Monuments Record, Sites and Monuments Records, Historic Environment Records)
  - ○ libraries and museums
  - ○ published material (e.g. historical and trade directories, construction texts, topographical studies)
  - ○ archive material (e.g. title maps, deed plans, rate books, census returns, local estate archives)
  - ○ graphic material (e.g. drawings, paintings, engravings, postcards (see Fig. 5.3), terrestrial and aerial photographs)
  - ○ cartographic material (e.g. Ordnance Survey sheets, geological maps[1], *Digital Geological Map of Great Britain*,[2] *Lanmark Information Group Limited*[3])
  - ○ on-line support services (e.g. *Images of England*[4])
- **Documentation**
  - ○ design and procurement documentation (e.g. permissions and consents, contracts, drawings, schedules, specifications, minutes of meetings, correspondence)
  - ○ reports (e.g. surveys, health and safety files, access audits, heritage impact assessments)
  - ○ agreements and guarantees (e.g. fire and intruder detection systems, damp-proofing, timber treatment, underpinning)
  - ○ agents' particulars
  - ○ documents of ownership (e.g. freehold, leasehold, tenancies, mortgages)
  - ○ maintenance and service charge agreements
  - ○ maintenance and operating manuals
  - ○ legal documents (e.g. details of covenants, easements, rights of way, rights to light, party wall awards, dangerous structure notices, closing orders)
  - ○ utility service bills
  - ○ listed building and scheduled monument descriptions and curtilage
  - ○ conservation area boundaries
  - ○ descriptions and map for sites included in the Register of Historic Parks and Gardens of Special Historic Interest in England
- **Oral and anecdotal information**
  - ○ owners and occupiers (past and present)
  - ○ building staff (e.g. cleaners, caretakers)
  - ○ grounds staff (e.g. gardeners)

Notes

1. In *Cormack & Another v Washbourne* (Court of Appeal, 6 December 1996) it was determined that the surveyor should have consulted a geological drift map to ascertain the likely soil type beneath a property with numerous cracks that had been identified during a survey.
2. The *Digital Geological Map of Great Britain* (DiGMapGB) project is currently preparing 1:50 000 scale data for England, Wales and Scotland, with information arranged in up to four themes – bedrock geology, superficial deposits, mass movement and artificial ground.
3. Landmark Information Group Limited provides an account of land use for the period *c*. 1840 to *c*. 1995 using historical maps and land use datasets.
4. The *Images of England* initiative is run by the National Monuments Record, the public archive of English Heritage, and aims to create a 'point in time' photographic record of every listed building in England.

Fig. 5.2 Primary and secondary sources of information.

Fig. 5.3   Postcard showing damage caused by lightning at Grove Farm, Terrington St John, Norfolk in April 1913. (Courtesy of Mr and Mrs C. Cousins.)

Report writing

The purpose of a report is to bring together all relevant information derived from the preliminary site visit, background research, and inspection or survey, and 'communicate to the client the implication of the building's condition' (Hollis, 1998, p. 8). The importance of a report, both as a record of condition and as a guide for decisions or actions, cannot be understated.

A report based on a building survey might typically include the following (Fig. 5.5).

The relevance of what is seen during a survey, and the consequences for the owner or client, must be carefully considered, as it is not enough simply to describe the defects without explaining their significance. Assessing the risk of a particular defect will therefore require a thorough understanding of the building, and of the needs and expectations of the client. Options will therefore need to be considered and evaluated, ranging – in the case of a single defect – from simple palliative measures (low cost, high risk) to complex repair or replacement (high cost, low risk).

For certain complex or important buildings it may be necessary to bring together a team of experts, who will inspect, assess and report on individual parts under the control of project leader. Historic structure reports, which are an American initiative based on a detailed interdisciplinary survey of an individual historic building prior to its reuse or refurbishment, offer a

- **Exterior**
  - roof coverings
  - parapet walls
  - roofscape features (e.g. cupolas, dormers, balustrading)
  - chimney stacks and flues
  - flashings and weatherings
  - plant rooms
  - gutters and rainwater pipes
  - foundations
  - main walls
  - structural frames
  - balconies
  - external escape stairways
  - damp-proof courses
  - subfloor ventilation
  - doors and windows
  - glazing

- **Building services**
  - water supply
  - foul-water disposal (e.g. septic tanks, cesspits)
  - rain and surface-water disposal (e.g. soakaways)
  - utility services
  - service fittings (e.g. sanitary fittings)
  - heating installations (e.g. fuel storage tanks)
  - mechanical services
  - telecommunications, computer networks and data links
  - cable and satellite television
  - fire protection systems
  - security systems
  - lightning protection

- **Interior**
  - roof structure and void
  - ceilings
  - walls and partitions
  - beams and columns
  - fireplaces, flues and chimney breasts
  - floors
  - cellars and vaults
  - doors and windows
  - glazing
  - stairways and ramps
  - internal joinery
  - thermal insulation
  - finishes and decorations
  - fixtures and fittings
  - health and safety (e.g. hazardous materials)
  - fire precautions

- **Site and environment**
  - access and circulation (e.g. roads, drives, paths, steps)
  - water courses
  - boundaries (e.g. walls, hedges, fences, ditches)
  - retaining walls
  - hard and soft landscaping
  - flora and fauna (e.g. trees, shrubs, climbers)
  - outbuildings
  - garden structures (e.g. fountains, grottos)
  - statuary and ornamentation
  - site security
  - disabled access
  - environment (e.g. pollution)

Fig. 5.4 Format of a typical building survey.

useful model for understanding the history, fabric and needs of a particular building (Slaton *et al.*, n.d.). The typical content of such a report is shown in Fig. 5.6.

A similarly interdisciplinary approach is taken in the preparation of conservation plans, which document the significance of particular buildings or sites and how that significance might be retained in any future use, alteration, repair or development (Fig. 5.7). A conservation management plan would additionally set out specific actions or proposals for the management of the heritage asset (Clark, 2001).

- **General information**
    - name and address of clients
    - property address and/or Ordnance Survey national grid reference
    - date(s) of survey
    - weather conditions at time of survey
    - purpose and scope of survey
    - tenure and occupancy
- **General description**
    - description of property
    - accommodation
    - outbuildings and parking
    - approximate age
    - orientation
    - location and amenities
    - summary of construction and materials
- **External and internal condition (incl. building services and site)**
    - general condition
    - deterioration in relation to age
    - description of defects
    - appropriateness of use
    - adequacy of maintenance and general care
    - major problems and defects
- **Further advice**
    - specific requirements (e.g. fire precautions, security precautions)
    - need for further investigations or monitoring
    - matters to be checked by legal advisers
    - matters that might materially affect value
    - recommendations and priorities for work
    - building insurance
    - open market value

Fig. 5.5 Typical content of a survey report.

Such detailed approaches to understanding a building can obviously be applicable only for those of particular importance or worth, but it does provide a format that might usefully be adapted and adopted for other needs. In assessing a building, it is this breadth of information that is needed in order to understand how it works and responds to its environment and usage.

Assessment of defects

The assessment of defects is typically undertaken through a series of on-site procedures that provide the basic information on which to make an assessment as to condition and fitness for purpose. These procedures require the surveyor to:

- **Methodology**
  - ○ procedures followed in executing study and report
- **History**
  - ○ narrative history of the site, building or structure
  - ○ identification of significant individuals associated with the property
- **Site and landscape evaluation**
  - ○ archival and physical research of the site and landscape
  - ○ evolution and condition of site and landscape
- **Archaeological evaluation**
  - ○ identification and condition of archaeological features
  - ○ contribution to understanding of site history and architecture
- **Architectural evaluation**
  - ○ archival and physical research of the structure
  - ○ chronology of construction
  - ○ documentation of missing features
  - ○ identification of original fabric
  - ○ condition of existing features and components
  - ○ room-by-room description
- **Structural evaluation**
  - ○ archival and physical research of structural systems
  - ○ evolution and condition of structural systems
- **Building systems evaluation**
  - ○ archival and physical research of mechanical, electrical, plumbing, fire and security systems
  - ○ evolution and condition of building systems
- **Construction chronology**
  - ○ evolution of structure
  - ○ physical and documentary evidence
- **Materials conservation analysis**
  - ○ primary building materials
  - ○ characteristics and composition
  - ○ interpretation of field and laboratory analysis
  - ○ identification of material failures
- **Identification of significant features**
  - ○ identification of features of note for historical significance, architectural and engineering design, materials or craftsmanship
- **Recommendations**
  - ○ cost estimates
  - ○ outline scope of work
  - ○ recommendations and alternatives
- **Bibliography and reference sources**
- **Appendices**
  - ○ drawings
  - ○ photographs
  - ○ copies of reference documents

Fig. 5.6 Typical content of a historic structure report.

- **Introduction**
  - ○ nature of the site
  - ○ previous work on the site
  - ○ designations
  - ○ reason for project
  - ○ limitations on work
- **History**
  - ○ historical research and documentary evidence for construction, use and alteration
  - ○ wider historical, artistic, architectural, landscape, industrial, technical or other context
  - ○ identification of significant individuals associated with the property
- **Description**
  - ○ description of building or landscape today
- **Development**
  - ○ results of analysis of change over time
  - ○ phasing to create an integrated narrative
- **Statement of significance**
  - ○ summary of importance of site or findings
- **Impact assessment**
  - ○ assessment of impact of any proposal
  - ○ suggestions for mitigating adverse impacts
- **Recommendations**
  - ○ Suggestions for further research and analysis
- **Bibliography, archive and sources**
  - ○ bibliography of maps and illustrations
  - ○ reference to previous work on site
- **Glossary**
  - ○ unfamiliar technical terms
- **Appendices**
  - ○ other relevant reports, detailed text, extracts from key relevant historical texts

Fig. 5.7 Typical content of a conservation plan (Clark, 2001, p. 99).

- inspect the defect closely
- record the defect by description, measurement, photograph or sketch drawing
- inspect and examine the construction around the fault for other indications of the problem (or related defects) – look internally and externally and at adjacent elements, and check if other hidden parts of the building could be affected
- ascertain as accurately as possible the exact form of the construction
- examine construction drawings/specifications if available for detailed information on the fabric of the building
- test defect if applicable (e.g. moisture content, relative humidity, take samples for analysis)
- inspect hidden areas (e.g. non-destructive survey)

- examine maintenance manuals if available
- open up the structure if required
- discuss the problem with the occupants of the building – when was the defect first noticed and is it getting worse?

Off-site procedures may include:

- reference to relevant published information (e.g. BRE reports and digests, Agrément certificates, TRADA publications, British Standards)
- consultation with specialists (after taking further instructions from client)
- commissioning more detailed examination/testing by specialists (after taking further instructions from client)

In addition to inspecting and assessing the condition of the physical structure of the building, other matters may need to be considered and reviewed:

- fire precautions
- public health requirements
- disability access
- health and safety issues
- garden or landscape features
- environmental matters
- scope for alteration or conversion
- conservation area, listed building and scheduled monument concerns

The detailed analysis of a building, and in particular its structural form, is time-consuming and costly. It is, however, essential for developing an understanding of a particular building and is thus often undertaken for important historic properties. Such an analysis may require detailed monitoring or testing, absolute and relative measurement, graphical representation and numerical calculation in order to help the surveyor in the diagnosis and prognosis of defects. Being able to monitor and evaluate the structure and fabric of an entire building over time is a goal yet to be achieved, but without such detailed information it is impossible to fully understand the nature of complex defects and predict the consequences of particular remedial actions.

Severity of defects

The severity of a particular defect or level of damage or decay that affects a building or part of its construction is typically assessed with regard to an

individual or prescriptive standard. Such a standard will typically take into account a number of issues depending on the type, use and importance of the affected building.

Severity of the defect

The severity of a particular defect will be judged against the effect that it has, and will have, on the structure, fabric, services and facilities of the affected building. This may be based on a pre-defined scale, such as that used to classify cracking in masonry (BRE Digest 251, 1995), or relate to specific issues associated with the occupancy and use of the building.

In the case of leases, it is breaches of repairing or other covenants, identified by means of an interim or final/terminal schedule of dilapidations, that form the basis for remedial works. The standard of such repairs will typically have regard to the age, character and locality of the premises (Proudfoot v Hart (1890) LR25 2 BD). Establishing whether a defect breaches a repairing covenant requires there to be actual damage (Calthorpe v McOscar (1924) 1 KV 716 CA); Acre Close Holdings v Silva Court Residents' Association, unreported, 1996); deterioration from a previous condition (Credit Suisse v Beegas Nominees Ltd. (1994) 1 EGLR 76; Post Office v Aquarius (1987) 281 EG 798), and damage to the subject-matter of the covenant (Quick v Taff Ely Borough Council (1985) 3 AllE 321).

Effect and consequences of the defect on the well-being of the building

The effect(s) that a defect will have on a building if it is not remedied will be dependent on various factors, including the nature of the defect, the occupation and use of the building, and future maintenance cycles.

Cost of remedial works

The cost of remedial works will be affected by various issues including the nature and extent of the defect; the ease with which the work can be undertaken; standards of materials and workmanship; how the work is to be planned, managed and implemented; and the requirements of the building owner/user and relevant statutory authorities.

Effect of visible defect on appearance, confidence and saleability

What might be considered to be a minor defect that required no direct action could, in the case of a prominent public building, necessitate immediate

remedial works to sustain the appearance and/or confidence of the public in the building and its user. Similarly, a building that is to be disposed of would typically have all obvious defects attended to, regardless of their severity. The quality of the remedial works in such cases might, however, be lower than expected under more usual circumstances.

Prioritising defects and remedial works

The priority given to remedial works in response to a particular defect will similarly depend on various issues and be categorised under headings such as 'urgent', 'necessary' and 'desirable'. These categories will dictate when the work is undertaken, and allow future works to be programmed and financed as part of a rolling programme of planned works.

Assessing the severity of defects and the corresponding priority for remedial works will typically include consideration of:

- statutory obligations
- health and safety issues
- functional and operational requirements
- lease or covenant obligations
- rate of deterioration and decay
- cost fluctuations
- value and utility of building and its facilities
- desired or expected standards
- running or operational costs
- management of risks

Unoccupied buildings and sites

Buildings and sites become empty for a variety of social and economic reasons. Some of these buildings may be considered beyond economic repair or adaptation, and the most practicable option will generally be complete or partial demolition, and the redevelopment of the site. For the vast majority of unoccupied premises, however, the best solution will be reuse, but until such time as a suitable use and occupier can be found, the property must be properly managed (Swallow, 1997).

A sensible approach to the inspection and assessment of such empty properties is required in order to formulate and implement an action plan to protect and maintain them in an appropriate condition suitable

Fig. 5.8   Unoccupied buildings with boarded lower windows, missing lead flashing to bay window, broken glass, pigeons inside upper rooms, broken downpipes, vegetation growth to frontage and guttering, and slipped roof slates and tiles.

for future reoccupation. Without proper management, unoccupied buildings will deteriorate and become the target for theft, vandalism and arson. Apart from creating a potential eyesore, this will have implications for the building owner and wider implications for the environment (Fig. 5.8):

- a potentially useful asset is not being exploited
- a building in disrepair has a negative effect on the surrounding community and environment
- an unoccupied building or site may present a health and safety hazard
- the building may deteriorate beyond economic repair leading to demolition and the possible loss of a building with townscape and/or architectural value

Building owners also have certain legal obligations and, in particular, owe a duty of care in tort to others (public, visitors, trespassers) who may suffer injury arising from acts or omissions in respect of nuisance, defective premises or negligence.

Redundant and ruined buildings

Whilst it is perhaps unnecessary to dwell too long on the subject of re-dundant and ruined buildings in a book dealing chiefly with buildings in use, it is nonetheless important to acknowledge and accept the differing conditions and requirements of those without current purpose or use.

Standing ruins are typically the remains of buildings or structures that have suffered natural or man-made action, and as a result can no longer be considered fit for occupation. As such, they may be thought of as 'dead' (belonging to a past civilisation or serving obsolete purposes) rather than 'living' (continuing to serve the purposes for which they were originally intended).

Despite being without 'life', ruins may have inherent value and so be worthy of preservation. This value may be derived from the particular ar-chitectural, archaeological or historical importance of the ruin, or from the part it plays, whether as an actual ruin or one constructed for the purpose, in the setting of a specific building or landscape (Fig. 5.9) (see Chapter 6).

Redundant buildings are not, by contrast, 'dead', but instead may be thought of as merely 'resting'. This assumes that they will regain their orig-inal purpose or use, or be adapted to fulfil some other need. The conversion and adaptive reuse of redundant buildings has long been a challenge for owners and their advisers, and is again receiving renewed interest in the context of economic development and sustainability (see Fig. 7.14).

Whether ruined or redundant, such buildings are typically exposed to specific and often accentuated levels of damage and deterioration. It is therefore important to acknowledge this in relation to present and future expectations.

In the case of a standing ruin, the structure and fabric of the building is often without the full protection usually provided by roofs and rainwater disposal systems. As a result it is the tops of walls, internal partitions, floors, ceilings and other finishes that are exposed to wind and rain, and which will inevitably suffer (Fig. 5.10). Inspection and assessment of such a ruin will therefore need to take account of:

- location and setting (e.g. importance in landscape)
- site features (e.g. water table, water courses, surface-water drainage, livestock, wildlife, vegetation, below-ground remains and voids)

Fig. 5.9   The love of natural and man-made features within a contrived landscape provided an ideal outlet for follies, eye-catchers and other idiosyncratic structures, such as at Wimpole, Cambridgeshire (*c.* 1772), during the eighteenth century.

- site activities (e.g. ploughing, grazing, vegetation clearance, public access)
- environmental conditions (e.g. exposure to wind, rain, atmospheric pollution)
- structure and fabric (e.g. structural assessment, material degradation and erosion, vandalism, theft)
- interpretation (e.g. history, causes of abandonment and ruination, surviving features, analysis, building phases, previous alterations and repairs, re-use of materials)
- health and safety (e.g. falling masonry, collapse, need for temporary works)

In the case of redundant buildings, there are further and wide-ranging considerations to consider. Such buildings may have a potential value, but are often simply abandoned to neglect and misuse (Fig. 5.11). Inspection and assessment of such buildings will therefore need to take account of various considerations, many of which have been noted for unoccupied buildings and sites above.

Fig. 5.10   Exposed wall heads and internal surfaces of Wingfield Manor, Derbyshire. Built in the mid-fifteenth century, this country mansion has been unoccupied since the 1770s.

Diagnosis and prognosis of defects

Defect diagnosis

Defects may be obvious and readily determined on examination of the building or particular element of construction, or be present in such a form or location that simple detection is not always possible. Hidden defects may remain undetected for many years, and cause serious damage. The detection and diagnosis of such latent defects requires all the skills of the surveyor and may require detailed investigation including the use of non-destructive survey techniques, material testing, measurement, monitoring or opening up of the building fabric.

The various indicators of building defects are as follows:

- visual (e.g. staining, cracking)
- physical (e.g. structural failure)
- olfactory (e.g. odours)
- aural (e.g. water hammer)
- tactile (e.g. uneven surfaces)

Fig. 5.11   Redundant industrial buildings on River Don in Sheffield.

Once a defect has been detected, it is crucial that its cause is fully understood and correctly diagnosed before any remedial action is taken. Diagnosis therefore requires collection and assimilation of all relevant information (see above), and is reliant upon:

- a knowledge of the behaviour of relevant building materials
- a knowledge of the construction of the building
- a knowledge of how the use (past, present and future) of the building might affect the construction

Each defect should be assessed on the basis of a logical series of questions and answers, posed initially during the survey and, if necessary, after research and further consideration (Fig. 5.12). Questions should seek to confirm or deny a single or combined cause for what is seen or inferred, and attempt to diagnose the defect through its relationship with the structure, fabric or environment within and around the building.

Where evidence of a potential defect is available to a surveyor by way of symptom or assumption, the probability of the defect being actually present must be assessed by considering all available information. The time spent in raising this probability, often through a process of elimination, will

- **Symptoms**
  - How does the defect manifest itself?
  - Do the symptoms change (e.g. with weather conditions) or are they constant?
  - Are the symptoms getting worse?
- **Investigation**
  - What is the extent of the defect and could it affect other parts of the building?
  - Are the symptoms relevant to one or more possible defects?
  - Could the cause of the defect be remote from the symptom?
- **Diagnosis**
  - What is the cause of the defect?
  - Can the defect be attributed to two or more causes?
  - Is the defect static or progressive?
  - Is further action required to diagnose the defect?

Fig. 5.12 Typical questions posed during defect diagnosis.

be reflected in time, resources and cost, and should be carefully considered with regard to the purpose of the survey and agreed brief.

Where such symptoms are complex or conflict with other available information, the surveyor may need to adopt a more scientific approach to diagnosis. In this, a particular line of reasoning or investigation is tested against known or assumed facts so that it might be proved or rejected in favour of another (Fig. 5.13). Such an approach is based on a number of stages:

- observations – observe and record visible symptoms
- theories – develop a number of theories based on observations
- questions – define a set of questions to differentiate between theories
- tests – find answers to questions (e.g. specialised survey, monitoring or chemical testing)
- analysis – analyse and interpret results
- conclusions – match diagnosis to symptoms and results
- feedback – redefine questions as required to support new information and prognosis
- action – develop appropriate course of action (e.g. repair, replace)

An example of applying this systematic approach to defect diagnosis is given in Fig. 5.14.

Although such an approach is based on standard questions and answers, it requires the knowledge and experience of the surveyor to observe all relevant symptoms and draw the correct conclusions from investigations or test results. It may also require the surveyor to mentally 'deconstruct' the building in order to assess a defect in the context of a particular element or material, and then to 'reconstruct' the building in order to make a prognosis based on an overall view of the building and its various systems.

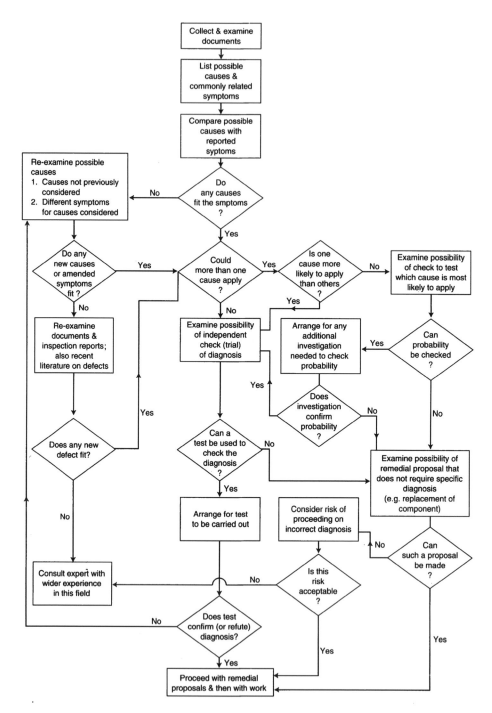

Fig. 5.13 Decision tree for the diagnosis of defects, based on that presented in *Defects in Buildings* (Property Services Agency, 1989, p. 22). © Crown copyright material is reproduced with the permission of the Controller of HMSO and Queen's Printer for Scotland.

- **Observations**
  - ○ Low-level staining to internal surface of outside wall in occupied house
- **Theories**
  - ○ Cosmetic, rising damp, penetrating damp, condensation, leakage
- **Questions – general**
  - ○ What is the room used for (e.g. bathroom, kitchen)?
  - ○ How is the house/room used (e.g. regularly, intermittently)?
  - ○ What is the construction of the wall (e.g. solid, cavity, timber frame)?
  - ○ What is the wall constructed of (e.g. bricks, blocks, damp-proof course, cavity insulation, plasters, renders, decorative finishes)?
  - ○ How is the room heated (e.g. paraffin heaters)?
  - ○ Does the staining indicate a current or past (and remedied) defect?
  - ○ Are there any obvious causes (e.g. bridged damp-proof course, defective pointing, leaking pipes)?
  - ○ Are there any concealed causes (e.g. faulty cavity tray, defective cavity tie, damaged vapour barrier, displaced insulation)?
  - ○ Are there specific indicators present (e.g. efflorescence, mould growth)?
- **Questions – specific**
  - ○ Is there a discernible pattern to the dampness (e.g. regular line, specific location)?
  - ○ Does the pattern of dampness match any features (e.g. raised flowerbed, abutment of garden wall or buttress)?
  - ○ Does appearance of staining relate to weather patterns (e.g. appears after heavy rain)?
  - ○ Is there adequate ventilation within the room?
- **Tests**
  - ○ Moisture levels (e.g. moisture or carbide meters, gravimetric analysis)
  - ○ Efflorescence (e.g. salt analysis)
  - ○ Environmental conditions (e.g. ambient and surface temperatures, absolute and relative humidities)
  - ○ Surface and/or interstitial condensation (e.g. calculation, environmental monitoring)
- **Analysis**
  - ○ Do test results indicate a specific defect?
  - ○ Are tests appropriate (e.g. moisture meters are usually calibrated to give readings of free moisture (%) in timber only or wood moisture equivalent (%) for other materials)?
  - ○ Could results be caused by other factors (e.g. high meter readings due to conducting materials or hygroscopic salts)?
  - ○ Should other tests be undertaken (e.g. moisture meter with deep-wall probes)?
- **Conclusions**
  - ○ Presence of moisture gradient across wall with peaks in corners indicative of rising damp
  - ○ Presence of individual damp patch indicative of specific addition/omission/defect of construction
  - ○ Presence of specific salts (e.g. nitrates and chlorides) indicative of rising damp
- **Feedback**
  - ○ Does the diagnosis remain valid in the presence of additional information (e.g. anecdotal information relating to blocking of former doorway now covered by render)?
  - ○ How is the wall likely to be affected (e.g. salt contamination of plaster, risk of fungal decay to adjacent and embedded timbers)?

Contd.

- **Action**
  - ○ What immediate action can be taken (e.g. lowering of raised flowerbed)?
  - ○ What repairs are appropriate (e.g. cutting out and replace decayed bricks, repointing open/defective joints, replastering, redecorating)?
  - ○ What alterations or improvements are appropriate (e.g. installation of damp-proof course, fitting bellmouth drip to lower edge of render)?
  - ○ What preventive maintenance should be undertaken and when (e.g. keeping air bricks clear, redecorating render every five years)?

Fig. 5.14 Systematic approach to defect diagnosis.

Automation of these processes through the use of computer data loggers or expert systems can be appropriate for stock surveys, but should not be relied upon for surveys of buildings of individual complexity or peculiarity in construction or form. The potential use of computers for modelling and analysing buildings, and logging the effects of defects on structure, fabric, services and occupants, may, nevertheless, make future analysis and diagnosis more responsive to complex and interrelated defects. Regardless of the form or sophistication of computer model or simulation, such techniques are, however, reliant on the accuracy and validity of the data on which they are based.

Defect prognosis

The prognosis of defects requires consideration of various factors concerning the condition and use of the building, and, in particular, demands a detailed understanding of the cause and severity of the defect, and an awareness of how it might advance over time. As with diagnosis, the prediction or forecasting of defects and the consequences of neglect are assessed on the basis of a logical series of questions and answers, which are made with regard to what is seen or can reasonably be inferred during or following a survey (Fig. 5.15).

- Will the defect progress further?
- How rapidly will further deterioration take place if no remedial action is taken?
- Will the stability of the building or well-being of the occupants be seriously threatened?
- If so, when will the situation become critical?
- Are any immediate safeguarding works necessary (e.g. temporary works, hoardings, debris netting)?
- What are the short-, medium- and long-term consequences (e.g. structural stability, health and safety, finance, occupation) of doing nothing?
- What is the most appropriate cause of action (e.g. replacement, repair, maintenance)?

Fig. 5.15 Typical questions posed during defect prognosis.

- **Future of the building:**
  - Is the building or the part affected by the defect nearing the end of its useful life?
  - If so, is it worthy of retention?
  - If the building is not worthy of retention and is nearing the end of its useful life, is it appropriate to undertake temporary or palliative action rather than an expensive permanent repair?
  - Are major alterations or demolition planned that would render any repairs valueless?
  - Could the repair incorporate features not only to remedy the defect, but also enhance the performance of the element or component concerned (e.g. increased natural ventilation, thermal insulation)?
- **Needs of the occupants:**
  - Can occupants remain safely in occupation during works?
  - Will the works adversely affect activities or sensitive contents?
  - Will they need to be decanted and to where?
- **Client's resources:**
  - What funds are available for remedial works?
  - If funding is limited, will a cheap holding operation be appropriate until funds are available for a more permanent remedy?

Fig. 5.16 Additional questions on which to base defect prognosis.

As part of making a prognosis of a particular defect or series of defects, consideration must be given to the likely future of the building, the needs of the occupants and the availability of resources (Fig. 5.16).

Non-destructive survey techniques

Less than ten per cent of the fabric of a building will usually be available for direct observation and assessment during a survey, whilst the structure and other parts of the fabric remain hidden below ground, within the construction or covered by finishes and decorations.

In certain cases it is necessary to investigate beneath the surface in order to understand the construction and identify actual or potential defects. Aside from physically opening up parts of the building, it is possible to employ non-destructive techniques to provide a level of information, based either on direct penetration into the fabric or indirect/remote imaging techniques, that will aid assessment.

The term 'non-destructive survey' covers a collection of techniques that may be used to inspect or observe materials or elements of construction in place without causing alteration, damage or destruction to the fabric of the building (Hutton, 1991; Demaus, 1996; Livingston, 1996; Nappi and Côte, 1997; Williams, 1997; Ballard, 2001). Each technique is able to provide a specific set of measurements or data in response to known or suspected

conditions. This may be for the analysis of material properties, detection of hidden aspects of construction or sub-surface anomalies, or the evaluation of performance in use.

As with all forms of investigation, success relies upon the correct selection of the technique or combination of techniques that are most appropriate for the particular task, and skilled interpretation of the collected data in order to reliably inform the decision-making process.

Radiography

Penetrating X-rays or gamma rays are used to see inside a wide range of materials and elements of construction to detect discontinuities, cracks, voids, density variations, hidden details, foreign objects or changes caused by deterioration. Materials such as plaster and wood are easily penetrated to show embedded fixings and structural members, but with denser materials, such as masonry and concrete, a more powerful radiation source is required.

Surface penetrating or impulse radar

A pulse of radio energy transmitted into the ground or solid material along a predetermined survey line is reflected by internal surfaces and objects, identifying boundaries, thicknesses and defects such as cracking of stone or concrete (Fig. 5.17). Examples of its use with buildings include determining the material thickness, moisture content, and locating voids and embedded metalwork (Baston-Pitt, 1993) (see Chapter 6).

Microwave analysis

The strength and direction of projected energy that is reflected back from surfaces and changes within a material can record inconsistencies, discontinuities, faults and hidden details. This technique has been used successfully on mass constructions, such as ramparts and redoubts found in military architecture.

Thermography

The measurement of infrared energy, in the form of waves radiated from the surface of building elements, is commonly used to detect and quantify heat losses and temperature variations through roofs and walls. It can also

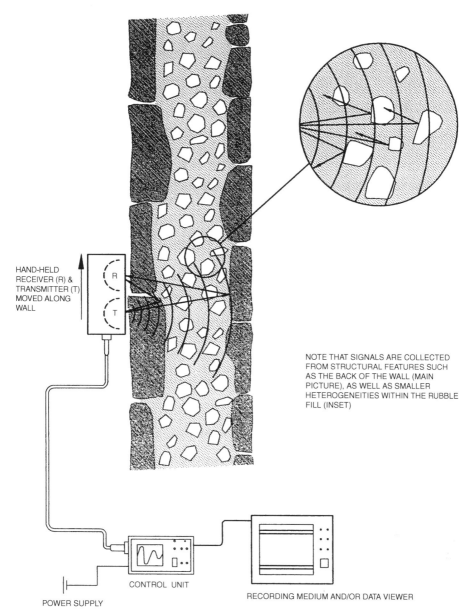

HAND-HELD
RECEIVER (R) &
TRANSMITTER (T)
MOVED ALONG
WALL

NOTE THAT SIGNALS ARE COLLECTED
FROM STRUCTURAL FEATURES SUCH
AS THE BACK OF THE WALL (MAIN
PICTURE), AS WELL AS SMALLER
HETEROGENEITIES WITHIN THE RUBBLE
FILL (INSET)

CONTROL UNIT

POWER SUPPLY

RECORDING MEDIUM AND/OR DATA VIEWER

Fig. 5.17   Basic theory of impulse radar, showing collection of data from within a rubble-filled masonry wall. (GBG.)

be used to detect sub-surface details, such as the presence of embedded timber-frame members, voids, inconsistencies, and deterioration in timber members. (BRE IP 7, 1990; BRE Report 176, 1991; McAvoy and Demaus, 1998).

Infrared photography

Based on the same technology as thermography, photography using in-frared film, together with appropriate filters, can indicate subtle variations in surface colour that might, for instance, result from an increased moisture content or sub-surface deterioration.

Acoustic testing

Various techniques have been developed for building and engineering applications that make use of the soundwave to detect inconsistencies within test materials. Sonic transmission and sonic echo testing uses transducers on both side of the material or structure to detect differences in time and frequency response.

Ultrasonic testing uses high-frequency soundwaves to detect and locate hidden details, voids, faults, cracks and cavities, and measure thickness and density (Demaus, 1992, 1995). This technique is particularly useful in dealing with homogeneous materials, such as ceramics, metals and masonry, rather than with less uniform forms of construction.

Magnetometry

The measurement of an induced magnetic field can, through distribution patterns, identify and locate magnetic features buried within non-magnetic materials such as wood, concrete, brick and stone. Cover meters are commonly used to locate and identify reinforcement within concrete constructions, but can also be used to detect metal cramps, armatures and fixings within masonry or timber (see Chapter 6), and buried pipes, conduits, flues and ducts. Certain non-magnetic materials and sub-surface anomalies can also be detected due to changes in magnetic fields induced by stimuli such as heat. Simple metal detectors are also widely available for use in the home to locate hidden cables, pipes and fixings.

Geotechnical surveying

The term 'geotechnics' describes an array of techniques and procedures used in investigating ground conditions, either in response to a defect or in advance of new development. These include destructive techniques such as sampling and testing using trial pits and boreholes, penetration testing, plate testing, as well as site and laboratory testing and monitoring of soils, rocks and contaminated landfill. Non-destructive techniques are also used, such as impulse radar for detecting cavities and obstructions below ground level.

Geophysical surveying

Archaeological prospection using non-destructive geophysical techniques can be used to detect sub-surface features that may have an importance in understanding a building or relevance to any planned works. Resistivity surveying is based on the resistance of an electric current passing through the ground relative to its moisture content. Sub-surface variations in this resistance can be interpreted to indicate features such as ditches, walls and built-up surfaces. Magnetic surveying, or magnetometry, is used to detect changes in the magnetised iron oxides in the earth caused by the fires and cultivation associated with occupation of a site. Magnetic anomalies can assist in the identification of industrial sites, hearths, kilns and filled pits.

Surface-penetrating radar is also being used to detect sub-surface features, often at depths of several metres, and is likely to become a more familiar technique as interpretation improves. Detection of sub-surface features can also come from direct probing, percussion variations known as bosing, by using metal detectors and dowsing.

Remote sensing

Remote sensing is the general name given to the acquisition and use of data about the earth from sources such as aircraft, balloons, rockets and satellites. Many of the techniques used in this field are applicable to the non-destructive surveying of buildings, and a number of photographic techniques have been employed in ground-based remote sensing to identify and record surface and sub-surface anomalies.

Rothound® search dogs

The detection and subsequent monitoring of dry rot (*Serpula lacrymans*) has traditionally been undertaken by periodic opening-up and direct inspection of suspect timbers. Rothound® dry rot search dogs are capable of detecting the scent of the metabolites produced by living dry rot, thus discriminating between living and dead outbreaks, and indicating the extent of an infestation (Hutton, 1991) (Fig. 5.18).

Gas sampling technology, probably in combination with micro-drilling (see below), may similarly provide a means of detecting fungal decay and possibly the presence of beetle infestations through the gases given off by faecal pellets.

Fig. 5.18   Rothound and handler during a survey. (Photograph by Hutton + Rostron.)

Dowsing

Dowsing can be used to discover buried features as well as its more common role in discovering underground water.

Close-circuit television surveying

Remote CCTV or video inspection and tracking of drain runs, flues, ducts and other concealed voids can provide direct information on condition, blockages and defects, and allow record drawings to be prepared for later use.

Micro-drilling

The rate of penetration of a fine drill probe (1 mm dia.) into timber can be used to identify and locate faults and variations due to decay and other defects, and inform later localised repair or treatment (Demaus, 1993, 1995) (Fig. 5.19). Entire structural members may be sampled by regular drilling

Fig. 5.19   Demonstration of micro-drilling.

and the condition of concealed timber members, such as lintels, assessed by drilling through overlying plaster.

Fibre-optic surveying

Direct penetration into cavities or voids within the fabric of a building using a fibre-optic probe can provide useful information on aspects of construction, or the condition of specific elements or components. The probes may be either endoscopes, which are flexible tubes filled with bundles of optical fibres that can illuminate and view remote objects, or boroscopes, which are rigid tubes, again with fibre bundles, that deliver illumination to the subject and carry the image back to the eye-piece. Both systems can be used to inspect and photograph hidden or obstructed details through existing cracks or openings, or small drilled holes.

Liquid penetrant testing

Application of a light oil impregnated with dye on to the surface of non-porous materials will provide a visual indication of cracks and fissures by the concentration of the dye in the defects. Dyed water (or smoke) may also be used to check pipe and drain runs, and also to detect leaks.

Load testing

Applying load to an individual member or assembly allows its behaviour to be assessed (Hume, 1998). The technique is useful in assessing damage following a fire or explosion, or establishing the efficacy of a repair. With this technique it is important to consider carefully the safety of the personnel undertaking the tests and the effects the tests might have on the building.

Monitoring defects

Most defects are progressive in nature and can be monitored over a period of time. In order to provide evidence as to the nature and timing of the defect, monitoring usually includes a level of absolute or relative measurement.

Early techniques of monitoring were simple, limited by the available resources and relatively undemanding user requirements. Increased sophistication has stemmed, in part, from a growing need to understand

and, where possible, quantify the cause of a building defect rather than just the effects it will have on the fabric.

In addition to monitoring current defects, it is important to appreciate the opportunities for comparing present defects with the state of the property in the past. In this respect the use of earlier photographs and other visual material, together with reliable proxy data, may assist in confirming and quantifying levels of decay, deterioration or disrepair (see Fig. 5.3).

How monitoring is performed depends on the difficulties of access, the overall condition of the fabric, the need to obtain a standard and continual sequence of information, and available finance. Accuracy will typically increase with complexity and cost. The reasons for monitoring must also be considered, whether it be to record progressive conditions, or detect certain changes or conditions before damage is sustained.

Structural distortion and movement

Monitoring of structural distortion and movement can take many forms, depending on the underlying cause and how it has manifested itself in the fabric of the building (Fig. 5.20) (BRE Digest 343, 1989, Digest 386, 1993 and Digest 344, 1995) (see Chapter 6). Cracking and in-plane deformation of

Fig. 5.20 Distortion exhibited by timber-framed buildings in Lavenham, Suffolk.

masonry can be undertaken using crack width gauges or rulers, calibrated tell-tales (such as those produced by Avongard™), vernier gauges and fixed studs, mechanical strain gauges (e.g. Demec™), or electronic sensors connected to data loggers.

The deflection of a structural member, which typically implies either flexure under increased load or movement as a result of material deterioration, can be monitored by levelling, inclinometers or electronic level gauges, whilst vibrations can be monitored by laser interferometry or with sensors and data loggers (Thomas *et al.*, 1997). Distortion of vertical surfaces, such as leaning or bulging walls, may be monitored using simple plumb-bobs or spirit levels, plumbed offsets or, more accurately, an optical plummet or by theodolite intersection.

Material deterioration and decay

Understanding how and why materials deteriorate and decay requires a detailed understanding of the particular material and the various agencies and mechanisms that can act upon it. Assessing the severity of such action with regard to defined criteria (such as aesthetics and structural considerations) may require initial and subsequent testing, whether on the actual material or an appropriate replicates, and monitoring to take account of varying conditions (such as time, temperature and humidity).

Tests have been established to determine key qualities and characteristics of common buildings materials (such as durability, porosity, water absorption and salt crystallisation), but it is also necessary to be able to assess performance in the context of a particular building, environment or application.

The diagnosis of particular defects may require monitoring over time, and this can be undertaken using various simple or sophisticated methods. Surveyors may, for instance, use specific diagnostic instruments, such as moisture meters, wall surface thermometers and salt detectors, to establish moisture levels in wood and other building materials, dew-point temperatures, the risk of surface condensation and the presence of hygroscopic salts, or undertake simple on-site tests to determine, for instance, moisture content (carbide meter), the presence of lead paint (lead test kit), the alkalinity of concrete (phenolphthalein indicator solution) or the presence of dry rot (Fugenex's *Dry Rot Sensors*®).

Generally, however, it will be necessary to use competent laboratory services in order to obtain data (qualitative and quantitative) to inform the diagnosis and prognosis of a particular defect or aspect of decay. Such services may include:

- laboratory and site testing
- diagnosis and analysis
- statistical analysis

Chemical and physical testing may be undertaken to identify particular materials, and measure levels of deterioration and decay. These include:

- timber (e.g. moisture content, preservative treatment)
- mortars (e.g. aggregate:binder ratios, cement content, aggregate grading)
- concrete (e.g. carbonation, strength)
- metals (e.g. composition, thickness of coatings, corrosion, fractures)
- asphalt and bituminous products (e.g. composition, bitumen content, hardness)
- glass (e.g. composition, surface discolouration)
- plastics (e.g. composition, compatibility)
- moisture analysis (e.g. carbide meter, gravimetric tests) (Kyte, 1997) (see Chapter 6)
- salt analysis (e.g. carbonates, chlorides, nitrates, sulphates)

Testing is also being undertaken at national and international levels to determine the responses of materials to varying environmental conditions. This includes the National Materials Exposure Programme (initiated in 1987), and its European counterpart, which is investigating the effects of pollution on building materials around the country (Butlin *et al.*, 1995) (Fig. 5.21). Such tests have been able to determine relationships between atmospheric conditions and the degradation of specific building materials, and so predict future patterns of material decay.

Research into the mechanisms of decay and treatment of historic building materials is also being undertaken by heritage organisations, academic institutions, and commercial partners, including projects related to climate change, flooding, and coastal maritime heritage, as well as magnesian limestone decay, and turf protection and wear (English Heritage, 2005).

Environmental conditions

The three environmental conditions that have the greatest effect on built fabric are temperature (ambient and surface), humidity (absolute and relative) and sunlight (lux and ultraviolet). Each can be measured and monitored to provide necessary information for corrective measures or suitable protection. Consideration may, in addition, have to be given to air movement and atmospheric pollution.

Fig. 5.21   Material testing as part of the National Materials Exposure Programme at Bolsover Castle, Derbyshire.

As with all monitoring, it is important to establish a pattern of readings over a period of time to take account of such variables as seasonal, diurnal and nocturnal differences; times and levels of occupancy; movement of caretaking, security and cleaning staff; and use of service installations.

On-site measurement may be undertaken using simple diagnostic instruments, but for more accurate measurement and monitoring the surveyor may be required to use or commission specific instrumentation:

- manual (e.g. thermometer, whirling hygrometer)
- mechanical (e.g. thermohygrograph)
- electronic (e.g. various humidity and temperature meters)

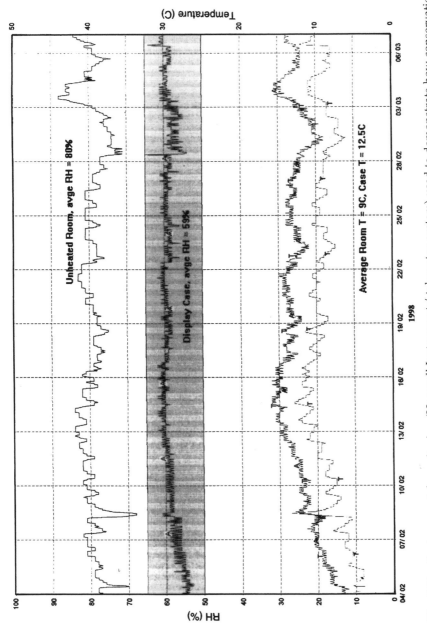

Fig. 5.22 Environmental monitoring (Hanwell Instruments' telemetry system) used to demonstrate how conservation heating can achieve a constant relative humidity (RH) inside a large display case within an unheated room. The RH-driven heating continuously adjusts the heating output to keep the RH at the chosen level as the ambient absolute air moisture content changes with weather cycles. In winter, a temperature uplift of 3–4 °C is all that is needed to maintain 60% RH, which lies within the 50–65% target range for mixed material collections. (Courtesy of Bob Hayes, Colebrooke Consulting Limited.)

- remote (e.g. sensors, data loggers and radio telemetry systems) (Fig. 5.22), including:
  - Grant Instruments' *Squirrel*[R]
  - Hanwell Instruments' *Humbug*[R]
  - Hutton + Rostron's *Curator*[R] (Hutton, 1995)
  - Gemini Data Logger's/*Tinytag*/*Tinyview*[R]

References

Ballard, G. (2001) *Non-Destructive Investigation of Standing Structures*. Technical Advice Note 23. Edinburgh: Historic Scotland.

Baston-Pitt, J. (1993) The role of impulse radar in investigating old buildings. *Construction Repair*, May/June, pp. 33–37.

Building Research Establishment (1989) *Simple Measuring and Monitoring of Movement in Low-Rise Buildings – Part 1: Cracks*. Digest 343. Garston: BRE.

Building Research Establishment (1990) *An Introduction to Infra-Red Thermography for Building Surveys*. Information Paper 7. Garston: BRE.

Building Research Establishment (1991) *A Practical Guide to Infra-Red Thermography for Building Surveys*. Report 176. Garston: BRE.

Building Research Establishment (1993) *Monitoring Building and Ground Movement by Precise Levelling*. Digest 386. Garston: BRE.

Building Research Establishment (1995) *Assessment of Damage in Low-Rise Buildings*. Digest 251. Garston: BRE.

Building Research Establishment (1995) *Simple Measuring and Monitoring of Movement in Low-Rise Buildings – Part 2: Settlement, Heave and Out-of-Plumb*. Digest 344. Garston: BRE.

Butlin, R., Yates, T. & Ashall, G. (1995) The United Kingdom National Materials Exposure Programme. *Water, Air and Soil Pollution*. **85** Part 4, 2655–60.

Clark, K. (2001) *Informed Conservation: Understanding Historic Buildings and Their Landscapes for Conservation*. London: English Heritage.

Demaus, R. (1992) Good vibrations (Ultrasonic testing of structural timber). *Context*, **35**, 28.

Demaus, R. (1993) Good vibrations II (Microdrilling of structural timber). *Context*, **39**, 25.

Demaus, R. (1995) Structural timber testing. *Construction Repair*, **9** (2), 23–27.

Demaus, R. (1996) Non-destructive investigations. *Building Conservation Directory 1996*. Tisbury: Cathedral Communications.

Dickinson, P. (1996) Building surveying. In *The Surveyor's Factbook* (P. Moreton ed.). London: Gee Publishing, Supplement 13.

English Heritage (2005) *English Heritage Research Agenda: An Introduction to English Heritage's Research Themes and Programmes*. Swindon: English Heritage.

Hollis, M. (1998) Report in full. *Valuer*, **67** (1), 8–9.

Hume, I. (1998) Load testing and historic structures. *Context*, **57**, 23–25.

Hutton, T. (1991) Non-destructive testing. *Building Research and Information*, **19** (3), 138–40.

Hutton, G. (1995) Building maintenance: The H & R 'Curator' and building monitoring systems. *Museum Management and Curatorship*, **14** (1), 92–110.

Kyte, C.T. (1997) *Laboratory Analysis as an Aid to the Diagnosis of Rising Dampness.* CIOB Construction Paper No. 80. Ascot: Chartered Institute of Building.

Livingston, R.A. (1996) Nondestructive materials characterization for historic conservation. *Materials Science Forum,* **210–213,** 751–8.

McAvoy, F. & Demaus, R. (1998) *Infra-Red Thermography in Building Survey and Recording: An Application at Prior's Hall, Widdington, Essex.* London: English Heritage.

Nappi, A. & Côte, P. (1997) Nondestructive test methods applicable to historic stone structures. In *Saving Our Architectural Heritage: The Conservation of Historic Stone Structures* (N.S. Baer & R. Snethlage, eds.), pp. 151–66. Chichester: John Wiley & Sons.

Property Services Agency (1989) *Defects in Buildings.* London: HMSO.

Slaton, D., O'Bright, A.W. & Hamp, P. (n.d.) *Guide to Preparation and Use of Historic Structure Reports.* Publication due in autumn 1999. Springfield: American Society for Testing and Materials.

Swallow, P. (1997) Managing unoccupied buildings and sites. *Structural Survey,* **15** (2), 74–79.

Thomas, P., Seeley, N. & O'Sullivan, P. (1997) The process of visitor impact assessment. *Journal of Architectural Conservation,* **3** (1), 67–84.

Williams, I.V. (1997) *Non-Destructive Testing of Historic Buildings.* Unpublished MA dissertation. Leicester: School of the Built Environment, De Montfort University.

Further reading

Association of Local Government Archaeological Officers (1997) *Analysis and Recording for the Conservation and Control of Works to Historic Buildings: Advice to Local Authorities and Applicants.* Chelmesford: Association of Local Government Archaeological Officers.

Ballard, G. (2001) *Non-Destructive Investigation of Standing Structures.* Technical Advice Note 23. Edinburgh: Historic Scotland.

Bowyer, J. (1979) *Guide to Domestic Building Surveys.* 3rd edn. London: Architectural Press.

British Standards Institution (1992) *Guide to the Durability of Buildings and Building Elements, Products and Components.* BS 7543. London: BSI.

Building Research Establishment (1990) *Structural Appraisal of Existing Buildings for Change of Use.* Digest 366. Garston: BRE.

Building Research Establishment (1991) *Surveyor's Checklist for Rehabilitation of Traditional Housing.* Garston: BRE.

Building Research Establishment (1991) *Why do Buildings Crack?* Digest 361. Garston: BRE.

Bussell, M. (1997) *Appraisal of Existing Iron and Steel Structures.* London: Steel Construction Institute.

Constable, A. & Lamont, C. (2006) *Case in Point – Building Defects.* Coventry: RICS Books.

Construction Industry Research and Information Association (1994) *Structural Renovation of Traditional Buildings.* Report 111. London: CIRIA.

Council for the Care of Churches (1995) *Guide to Church Inspection and Repair.* 2nd edn. London: Church House Publishing.

Douglas, J. (1995) Basic diagnostic chemical tests for building surveyors. *Structural Survey*, **13** (3), 22–27.

Fidler, J. (1980) Non-destructive surveying techniques for the analysis of historic buildings. *Transactions of the Association for Studies in the Conservation of Historic Buildings*, **5**, 3–10.

Harris, S. (2001) *Building Pathology: Deterioration, Diagnostics and Intervention*. Chichester: John Wiley & Sons.

Health and Safety Executive (1990) *Evaluation and Inspection of Buildings and Structures*. Health and Safety Series HS(G)58. London: HSE.

Hillam, J. (1998) *Dendrochronology: Guidelines on Producing and Interpreting Dendrochronological Dates*. London: English Heritage.

Holland, R., Montgomery-Smith, B.E. & Moore, J.F.A. (eds.) (1992) *Appraisal and Repair of Building Structures*. London: Thomas Telford.

Hollis, M. (1991) *Surveying Buildings*. 3rd edn. London: Surveyors Publications.

Hum-Hartley, S.C. (1978) Non-destructive testing for heritage structures. *Association for Preservation Technology Bulletin*, **X** (3), 4–20.

Institution of Structural Engineers (1980) *Appraisal of Existing Structures*. London: Institution of Structural Engineers.

Institution of Structural Engineers (1991) *Guide to Surveys and Inspections of Buildings and Similar Structures*. London: Institution of Structural Engineers.

Loss Prevention Council (1996) *Code of Practice for the Protection of Unoccupied Buildings*. 2nd edn. Borehamwood: Loss Prevention Council.

Melville, I.A. & Gordon, I.A. (1997) *The Repair and Maintenance of Houses*. 2nd edn. London: Estates Gazette.

Melville, I.A., Gordon, I.A. & Murrells, P. (1992) *Structural Surveys of Dwelling Houses*. 3rd edn. London: Estates Gazette.

Moore, J.F.A (ed.) (1992) *Monitoring Building Structures*. Glasgow: Blackie.

Noy, E.A. (1995) *Building Surveys and Reports*. 2nd edn. Oxford: Blackwell Scientific Publications.

Osborn, M. & Meeran, R. (1998) Warn home-buyers of the dangers of lead paint. *Chartered Surveyor Monthly*, **7** (7), 36–37.

Oxley, R. (2003) *Survey and Repair of Traditional Buildings: A Conservation and Sustainable Approach*. Shaftesbury: Donhead Publishing.

Robson, P. (2005) *Structural Appraisal of Traditional Buildings*. 2nd edn. Shaftesbury: Donhead Publishing.

Ross, P. (2002) *Appraisal and Repair of Timber Structures*. London: Thomas Telford.

Royal Institution of Chartered Surveyors (1991) *Surveying Safely: A Personal Commitment*. London: RICS.

Seeley, I.H. (1985) *Building Surveys, Reports and Dilapidations*. Basingstoke: Macmillan.

Shelbourn, M., Aouad, G., Hoxley, M. & Stokes, E. (2000) Learning building pathology using computers – A prototype application. *Structural Survey*, **18** (2), 111–19.

Staveley, H. & Glover, P. (1990) *Building Surveys*. 2nd edn. London: Butterworth-Heinemann.

Watt, D.S. & Swallow, P.G. (1996) *Surveying Historic Buildings*. Shaftesbury: Donhead Publishing.

Chapter 6
Remediation in Practice

Putting principles into practice

Having now considered building performance (Chapter 3), defects, damage and decay (Chapter 4), and the survey and assessment of buildings (Chapter 5), it is the intention of this chapter to bring together relevant information in order to plan and implement appropriate remedial action. In this, it is acknowledged that such action can be taken only once the building has been fully investigated, and so what follows is a series of case studies that seek to demonstrate the principles upon which this book is based.

The following case studies have therefore been chosen to demonstrate how investigative and remedial action has been designed, specified and implemented for specific situations involving the principal causes of defects, damage and decay used above.

Earthquake-resistant housing in Peru

Introduction

The development of an improved form of construction that offers greater resistance to earthquakes, whilst at the same time making use of local labour and materials, has been achieved by understanding the limitations of existing technologies and working together with those having to rebuild their homes and villages. Close collaboration and discussion have meant that the resulting buildings are closely suited to the needs of their owners, and offer an affordable and safer alternative to building with conventional masonry. This case study draws on work undertaken by Practical Action and presented in *Building in Partnership: Earthquake Resistant Housing in Peru* (Lowe, 1997).

Disaster and rebuilding

When, in May 1990, Alto Mayo in Peru was struck by an earthquake measuring six on the Richter scale, the effects were felt over 8000 square kilometres – damage was sustained by 8000 houses, with nearly 3000 irreparably affected. In two district capitals, eight out of ten houses were destroyed (Fig. 6.1).

Fig. 6.1   Tapial house in Soritor. (Courtesy of Practical Action.)

Development work between Practical Action, an organisation involved in achieving long-term sustainable development, and local groups sought to provide a means by which local people could rebuild their houses using local raw materials without reliance on expensive *materiales nobles* such as firebrick and reinforced concrete. These new houses had also to be designed and built to offer resistance to future earthquakes within the limitations of indigenous building technologies.

Developing an existing technology

Having studied the nature and extent of damage following the earthquake, it was seen that many of the *tapial* (rammed earth) houses had failed, but

those built of *quincha* (a type of wattle and daub) had shown a degree of resistance. The decision was taken that an improved form of quincha, which could be built with local materials and labour, would be the preferred construction to resist future earthquakes.

Quincha technology has been used in parts of Peru for many centuries, and the houses take the form of a round pole frame that is infilled with smaller-section timbers or bamboo and interwoven to form a framework that is plastered with one or more layers of earth. Development work undertaken by Practical Action sought to improve the performance of quincha through a number of changes (Fig. 6.2):

- concrete foundations to give greater stability
- wooden columns treated with tar or pitch and set into concrete
- concrete wall bases used to prevent moisture damage to timber wall construction
- improved jointing between columns and beams
- canes woven in a vertical lattice to offer greater stability
- lightweight metal sheet roofing used instead of roofing tiles
- tying of beams and columns with roof wires to improve performance in strong winds and earthquakes
- overhanging eaves to protect walls against heavy rainfall

Participation and success

A community building was constructed with improved quincha to raise awareness and promote greater understanding, and to allow individuals to learn through taking part in the construction process. Designs for new housing were also discussed with local people in order to develop a flexible approach that was adaptable to specific family needs. The result was a rebuilding project that engaged people in the whole construction process and provided affordable housing – the estimated cost (at 1996 prices) of a finished building of improved quincha was US$ 1299, whilst that built of brick using contract labour was US$ 5400.

An earthquake in April 1991 damaged a further 17 000 houses, but showed the improved resistance of those newly built with improved quincha. This has led to the wider adoption of this form of construction, with local variations to suit the availability of materials and needs of the users. The result is a comfortable, affordable and safe house that reflects the skills and individual priorities of the owners and builders (Fig. 6.3).

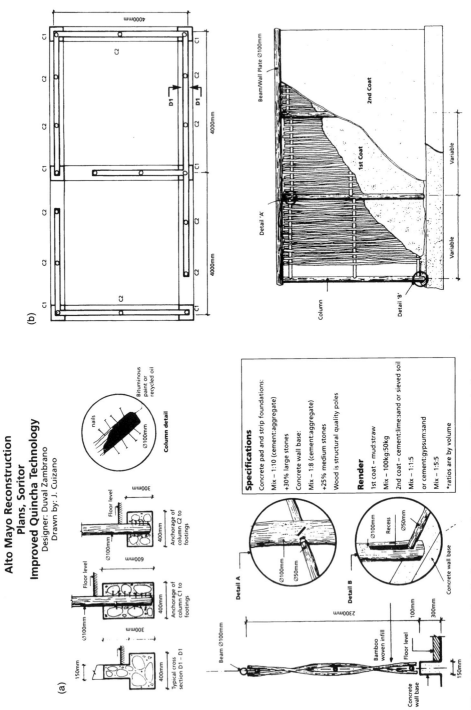

Fig. 6.2   Improved quincha technology. (Courtesy of Practical Action.)

Fig. 6.3   Improved quincha building in Jepelacio. (Courtesy of Practical Action.)

Stone deterioration by salt action at Walpole St Andrew, Norfolk

Introduction

The continual deterioration of stonework inside a redundant church has prompted detailed investigations into the mechanisms of moisture movement and salt crystallisation. By understanding the primary and secondary causes of these defects, it is intended that an appropriate level of remedial action may be taken to halt further decay. This case study draws on work undertaken by the author on behalf of the Churches Conservation Trust (Watt & Colston, 2000; Colston *et al.*, 2001).

Survey

The stone of the six arcade piers, being an oopelsparite limestone, is eroded and uneven, and the surface is loose and friable (Fig. 6.4). As a result, there is typically an accumulation of detached material, which combines stone dust and failed limewash, beneath each pier. The deteriorating condition of the stone has been a concern to parishioners and inspecting architects for most of the twentieth century, and various attempts have been made to reduce the continual deterioration. This included experimental desalination and

Fig. 6.4   Evidence of dampness and salt crystallisation to stonework of arcade piers.

consolidation of selected arcade piers, based on irrigation, poulticing and application of an alkoxysilane consolidant in the early 1980s.

Limewash has been applied to the stonework on various occasions, and the present application has become detached almost entirely from the piers and, to a lesser extent, from the corresponding responds. This detachment occurs above a relatively consistent height of 1.8 m to the north piers, and somewhat higher to the south, and extends up to the springing of the arches. Between the floor and the start of the failure, the limewash is generally sound and stable. Much of the limewash to the clerestory is also failing and becoming detached from a render finish.

A number of the horizontal joints to the piers, certainly at the higher and exposed levels, have been repointed with a hard cementitious mortar. Where this pointing has failed or been removed, it appears to be only

5–10 mm deep and overlies a soft lime-based mortar. Localised stone decay at a number of these joints has left the mortar slightly proud of the surface.

Moisture analysis

Drilled mortar samples were taken from mortar joints to the piers and responds, and from separate positions in the aisles to provide comparative data. The samples were gravimetrically analysed using a standard methodology for the measurement of moisture content and hygroscopicity, based on a comparison of sample weights as found, after exposure to 75% relative humidity and when oven dried (BRE Digest 245, 1981).

The moisture content profiles for the mortar samples show high values at floor level, typically in the range 12–17%, and progressively lower values at the higher levels. The hygroscopic moisture content profiles for the mortar samples show, by comparison, relatively low values at floor level, typically in the range 2–3%, but increase dramatically and progressively at the higher levels in the range 9–11%.

From these results it is clear that the moisture content in the lower parts of the walls and piers is dominated by rising damp, and that above this the moisture content is controlled by the hygroscopicity of the masonry (Fig. 6.5). The control samples taken from the aisles indicate the presence of rising damp, but do not show the presence of hygroscopic salts.

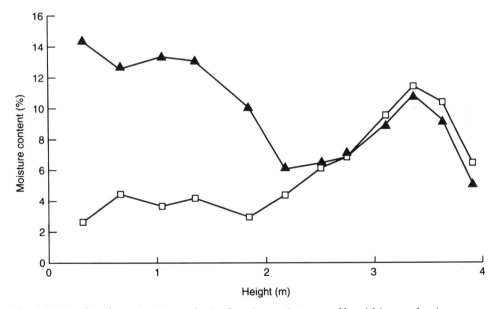

Fig. 6.5   Results of gravimetric analysis, showing moisture profile within arcade pier.

Salt analysis

Drilled stone samples were also taken from three locations to each of the piers and responds. The samples were analysed using ion chromatography and show a profile of sodium chloride salts within the piers. The presence of these salts is believed to relate to earlier contamination by seawater through flooding or as the result of surface treatment based on common salt undertaken at the end of the nineteenth century.

The cycle of dissolution and recrystallisation of the salts, which appears to be the underlying cause of the stone decay, is related to the environmental conditions within the church. Monitoring of internal ambient temperature (°C) and relative humidity (% RH) over a three-year period has shown high levels of relative humidity, typically in the range 70–80%. As sodium chloride goes into solution at about 75% RH, any fluctuations above and below this value will lead to damage to the stone. Extended periods of time (June–September) below 75% RH have been shown to lead to extensive splashing from the stone surface (Fig. 6.6).

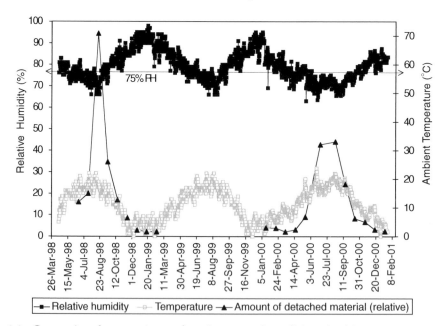

Fig. 6.6 Comparison between internal environmental conditions (ambient temperature and relative humidity) and the extent of deterioration of six stone piers.

Remediation

Previous investigations into the causes of the stone decay have likewise concluded that the principal mechanism is that of salt crystallisation. Remedial action has been attempted using the (then) experimental

technique of poulticing and alkoxysilane consolidation on two of the piers, although there is now no visible difference between those that were treated and the others in the arcades.

The design and implementation of remedial works will draw on what has previously been done and the results of the investigations outlined above. Such works will clearly have to take account of the high levels of moisture and salts within the piers, and the high levels of relative humidity within the church, but will also have to be appropriate for a church that is no longer in use for worship.

Metal corrosion and cathodic protection at the Inigo Jones Gateway, London

Introduction

The development and installation of a system of cathodic protection to retain wrought-iron cramps within the stonework of the Inigo Jones gateway in the grounds of Chiswick House required the use of non-destructive survey techniques for the location of the fixings and modification of an existing technology to provide a system of protection that required minimum intervention and a means of remote monitoring (Fig. 6.7).

Fig. 6.7   Inigo Jones Gateway. (Photograph courtesy of English Heritage.)

This case study draws on work undertaken by English Heritage and reported in *English Heritage Research Transactions – Volume 1: Metals* (Blackney & Martin, 1998).

Survey and diagnosis

A survey undertaken in 1990 identified three principal forms of decay affecting the stonework of the gateway – the formation of calcium sulphate skins as the product of the reaction between calcium carbonate and sulphur dioxide and trioxide in the urban atmosphere; localised decay due to the earlier use of hard repair mortars; and the corrosion of embedded wrought-iron cramps, causing widening of joints and cracking of surrounding stone.

Standard practice for the replacement of such cramps involves dismantling or cutting out sections of stone to allow for removal and replacement with non-ferrous fixings, followed by rebuilding or stone repair. Both approaches are intrusive and damaging. In the case of the gateway, the extent of dismantling would have been extensive and an alternative approach, based on an evaluation of risks with a preferred strategy for conservation based on sustained maintenance, adopted.

Remediation

A non-destructive survey of the gateway was undertaken using impulse radar to ascertain the number, position and condition of the iron fixings (Fig. 6.8). This technique is based on the transmission of pulsed radio energy, which is reflected back to a receiver where it meets changes within the structure. The resulting data need careful interpretation, and each fixing was finally located using a combination of impulse radar and high-resolution metal detection (including the development and use of a 5 mm diameter probe with unidirectional metal-detector head).

The technique of impressed current cathodic protection (ICCP), which is more commonly used for the rehabilitation of deteriorated steel in reinforced concrete, was chosen to halt the corrosion of the embedded cramps and so avoid the dismantling of the stonework. ICCP works by converting the metal to be protected into a cathode and moving the anodic reaction to a more durable metal. In the case of the gateway, the system comprised two-piece platinised titanium electrodes placed on either side of each cramp, with reference electrodes set at selected locations to determine the condition of the metal and enable the setting of the current and voltage. The installation of the system required the stonework to be accurately drilled and each cramp connected with plugs or self-tapping screws to ring main wires.

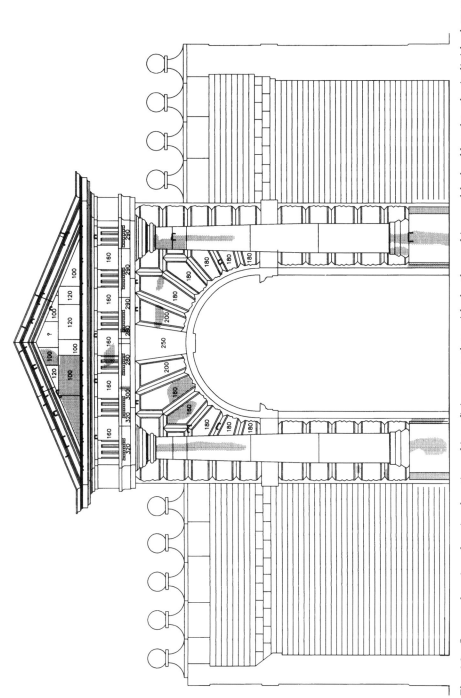

Fig. 6.8 Survey drawing showing location of iron fixings, together with depth, cracking and lack of bond to the individual stone blocks. (GBG.)

The ring main wires from the cramps, anodes and return leads from the reference electrodes are located in existing mortar joints and chases in the repair mortar render. The collected wires are run through an underground conduit to Chiswick House, where they connect to the power supply. Data loggers continuously record the output voltage and cramp potentials in relation to climatic changes and consequential cycles of masonry wetting and drying. The system might also be extended in the future to connect previously unidentified cramps.

Additional repair works included the provision of a new lead-sheet covering to the pediment of the gateway. This has been insulated from the stone or render to avoid ICCP currents initiating corrosion to the underside of the lead sheet.

Since the installation of the ICCP system there have been no instances of stone disruption due to cramp corrosion, which may be considered a successful outcome. Periodic monitoring of the system in 2001 revealed that the transformer rectifier that controls the power supply had failed and that a number of embedded reference electrodes were providing faulty readings. The opportunity was taken to modify the installation and a magnesium-powered remote sacrificial anode was inserted into the ground at the base of the gateway to provide current to the system. This work was tested over a period of 12 months and the results confirmed that the cramps had become polarised and were therefore well protected.

Chemical treatment residues at Melton Constable Hall, Norfolk

Introduction

The discovery of an unknown crystalline material during a survey of Melton Constable Hall and its subsequent identification as pentachlorophenol (PCP), a potentially hazardous substance used extensively for the treatment of fungal infections, required a detailed risk assessment to ensure that correct safety measures would be taken by those responsible for the building and for any future works that might take place within the affected rooms.

This case study draws on work undertaken by the author on behalf of the owner of Melton Constable Hall.

Survey and diagnosis

During a survey carried out by the author in August 1994, surface crystals were noted on the walls of four basement rooms (Fig. 6.9), and verbal

Fig. 6.9 Basement wall with crystals of pentachlorophenol following earlier irrigation treatment.

evidence given by the present occupant that persons entering these rooms were subject to bouts of sneezing.

The presence of these crystals was considered to be related to the treatment of dry rot (*Serpula lacrymans*), understood to have been undertaken during the late 1970s or early 1980s. Many of the walls within the property appeared to have been drilled and irrigated with a fungicide, and from contemporary texts the most likely treatments would have been based on sodium fluoride or magnesium silico-fluoride, sodium-PCP, sodium orthophenylphenate or a proprietary fungicidal solution based on PCP.

A sample of the crystals was analysed and the material identified as PCP (Cope *et al.*, 1995).

PCP was introduced to the market as a wood preservative in 1936 and became extensively used on a world-wide basis until the 1980s. Health concerns led to a reduction in use, and the ninth amendment to the European Economic Community marketing and use directive required the United Kingdom government to withdraw approval for the use of PCP after 1 July 1992. PCP was still allowed to be used under temporary relaxations for professional treatment 'in buildings of artistic, cultural or historic

interest or in emergencies to treat against dry rot fungus or cubic rot fungus', with notification to the Building Research Establishment. PCP is recognised as an existing active substance in the Biocidal Products Directive (98/8/EC), but is no longer supported for use.

The health risks associated with PCP are that it is toxic in contact with skin and if swallowed; very toxic by inhalation; irritating to eyes, respiratory system and skin; and has possible risks of irreversible effects. Self-contained breathing apparatus is required to be worn as toxic fumes are produced when the substance is involved in a fire. PCP has also been found to be carcinogenic in laboratory animal studies, and both PCP and sodium-PCP are suspected of possible mutagenic or teratogenic properties. Small levels of dioxins, which can cause poisoning and chloroacne, may also be present.

Having identified the residue as PCP and alerted the occupants to the potential health risks associated with the material, an assessment of risk was required for persons entering the rooms and undertaking any form of construction work.

Sampling and analysis

Samples of brick and crystalline residues from irrigation holes, surface dust, airborne particulates and indoor air were taken and analysed using gas chromatography and gas chromatography-mass spectroscopy.

Results and conclusions

The results obtained from analysis of the samples confirmed PCP to be present as a crystalline residue and as a contaminant of the brickwork within the sampled irrigation holes. This may be related to its movement within the basement walls in the presence of moisture or to its initial 'blooming' in the absence of a render finish. PCP was also present in dust samples taken from the floor and window sill, due probably to the detachment of crystals from the affected wall surfaces, whether by physical abrasion or air movement, and distribution onto surfaces close to the walls. There was no evidence for PCP within dust samples taken at high level within the room, which might typically have been associated with the volatile contamination of dust or suspended particulate material, and no volatile emissions within the room or in the room above.

The implications for those responsible for Melton Constable Hall, and for undertaking future works within the basement of the property, are

that the risks associated with the PCP-based treatment residue need to be considered and further assessed in relation to the individual tasks being proposed. Removal of surface residues and treatment of the affected wall surfaces, whether by containment and/or decontamination, is being considered. Long-term monitoring of conditions within the affected rooms should form part of any such works.

Engineering solution for the leaning tower of Pisa, Italy

Introduction

The famous leaning tower of Pisa, built as a campanile to the Cathedral between 1173 and 1370, has been the subject of many studies to understand and predict past and future movement. Recent research has been able to deduce the history of the inclination and assist in the development of appropriate stabilisation measures using an innovative technique of soil extraction. The technique is consistent with the requirements of architectural conservation and its implementation has required advanced computer modelling, large-scale development trials, an exceptional level of continuous monitoring, and day-by-day communication and control.

This case study draws on the work undertaken by the Italian Prime Minister's Commission for stabilising the leaning tower of Pisa and presented in a paper in the quarterly journal of the Royal Academy of Engineering in November 2001 (Burland, 2001).

Survey and diagnosis

The tower, which is constructed as a hollow cylinder with inner and outer facings of marble, is nearly 60 m in height, with foundations 19.6 m in diameter, and weighs 14 500 tonnes. At present the foundations are inclined at about 5.5° to the horizontal, with the average inclination of the axis of the tower somewhat less due to the slight curvature of the tower. As a consequence of this inclination, the seventh cornice overhangs the first by about 4.5 m (Fig. 6.10).

In order to understand the cause and effects of the tilting, it was first necessary to comprehend how the tower had been constructed and interpret the evidence drawn from many sets of measurements taken over the years. Careful observation of these records, together with detailed measurement of the masonry courses, revealed that the tower had already been tilting significantly during the period of building, and that adjustments

Fig. 6.10 The leaning tower of Pisa. (Photograph by John Burland.)

had been made when work recommenced after breaks in the construction process.

Computer modelling and analysis were undertaken to develop an understanding of the mechanisms controlling the behaviour of the tower, based on a simulation of its construction. Calibration of the model was varied to obtain agreement between actual and predicted present-day inclination. Conclusions drawn from this work showed that the instability of the foundation was not due to a shear failure of the ground, but rather the compression of underlying Pancone clay.

The inclination of the tower has also been shown to be very sensitive to ground disturbance, with past changes in inclination being related to specific events, including various remedial actions. A further discovery was that the tower was rotating about a point at the level of the first cornice, and that this could be related to seasonal differences in the level of groundwater.

Remediation

Having developed an understanding of the behaviour of the tower, it was agreed that stabilisation should be carried out in two stages. Temporary stabilisation of the masonry was undertaken by means of pre-stressed steel tendons positioned at various levels around the tower, having the effect of closing several of the cracks and reducing the risk of buckling failure of the marble cladding. Further temporary works were carried out on the foundations by applying a counterweight to the north side of the tower in order to reduce an overturning moment (Fig. 6.11). Precise measurements were taken throughout these operations, and these compared favourably with advance computer predictions of changes in inclination and settlement.

A permanent solution was sought that would result in a small reduction in inclination by half a degree – not enough to be visible, but sufficient to reduce the stresses in the masonry and stabilise the foundations. A method known as soil extraction was gradually evolved, which involved installing a number of soil extraction tubes adjacent to and just beneath the north side of the foundations (Fig. 6.12).

Following successful preliminary soil extraction in 1999, full soil extraction commenced in January 2000 using 41 extraction holes. By June 2001 the inclination of the tower had been successfully reduced by the intended half a degree. The temporary lead weights and pre-stressed steel tendons were removed. The seasonal differences in groundwater that were the cause of the continuing movement have been stabilised by means of radial drains leading to wells within which the water levels are fixed. In addition to reducing the inclination of the tower, a limited amount of strengthening work has been carried out on the most highly stressed areas of masonry. Having been closed to the public since January 1990, the tower was reopened in December 2001.

Understanding user requirements at the Greengate Medical Centre, London

Introduction

A building can be used successfully only if there is a close match between the needs of the user and the accommodation and facilities available. By studying these needs and understanding how a building is used, it is possible to seek improvement through the manipulation of internal spaces, whilst also creating an exciting living or working environment.

Fig. 6.11 Construction of north counterweight. (Photograph by John Burland.)

This case study draws on material provided by Narendra Gajjar, architect, and relates to a project undertaken on behalf of his clients, Dr Kalhoro and Dr Gopi. The Greengate Medical Centre opened in July 1997.

Survey and analysis

When a building is used for a different purpose from that for which it was built, there are often compromises that have to be made in terms of what the building can provide and the ways it can be used. In the case of the Greengate Medical Centre, the building had originally been designed and built as a house in the nineteenth century, and had been converted

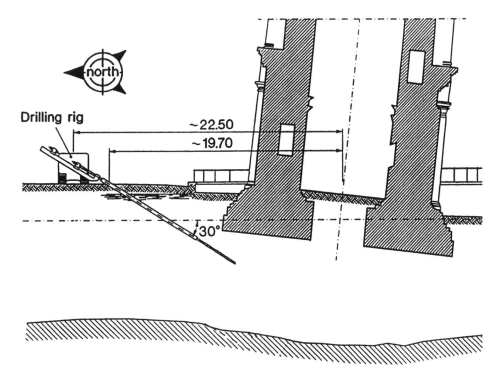

Fig. 6.12   Induced subsidence by soil extraction. (Photograph by John Burland.)

to a doctor's surgery in the 1960s. The surgery utilised the building without understanding how the various activities flowed between the different spaces and in a manner that therefore did not make the most of the available accommodation and site (Fig. 6.13a). As a result, the building was inefficient and patients were left confused by the internal layout.

Design principles

There are only a few ways in which a building can realistically be used for a particular activity without having to drastically alter the building or compromise the ways in which the activities are performed. Having analysed the limitations of the building and explored ways in which the spaces might be manipulated to meet the particular needs of the practitioner clients (such as the number and timing of clinics, and additional activities that may affect the proposed size of the waiting area), a set of clear design principles was devised that attempted to make best use of the building whilst creating a sense of interest and improved spatial awareness.

(a)

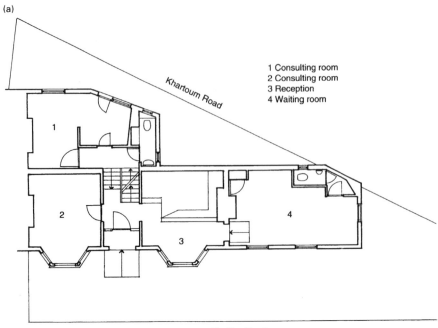

1 Consulting room
2 Consulting room
3 Reception
4 Waiting room

(b)

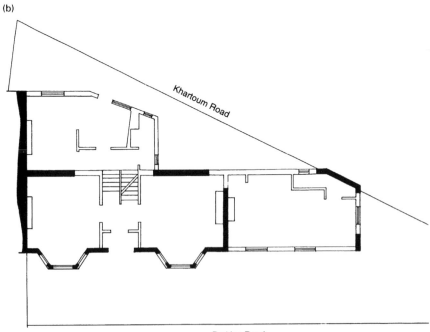

(c)

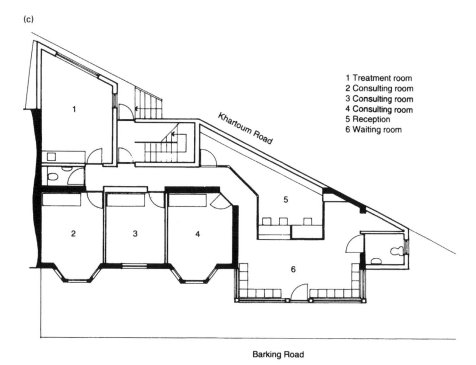

1 Treatment room
2 Consulting room
3 Consulting room
4 Consulting room
5 Reception
6 Waiting room

Khartoum Road

Barking Road

Fig. 6.13   Design principles. (Courtesy of Narendra Gajjar.)

These design principles had also to satisfy the requirements of the Department of Health and the East London and The City Health Authority in terms of the sizes and standards of accommodation. Further, a service development plan is prepared by the practitioners, which establishes the key facilities and priorities for their surgery, and determines what cannot be relinquished if the building is unable to accommodate all of their requirements.

Design solution

The design solution recognised the limitations imposed on circulation and space utilisation by the existing central stairway, and reordered the internal spaces to take account of a revised arrangement of structural walls and level changes (Fig. 6.13b). A new arterial corridor provides a link between the various rooms and spaces, and offers easier circulation for both staff and patients.

Opportunities were also taken to utilise the awkward end-of-terrace site by extending northwards up to the line of Khartoum Road, so providing additional space for a new stairway, treatment room and reception,

Fig. 6.14   Norton Park exterior. (Courtesy of Burnett Pollock Associates.)

and pushing the existing waiting room outwards to the south (Fig. 6.13c). Limited staff car parking was also fitted into the tapering eastern end of the site, by using a remote-operated electric sliding gate.

This new arrangement of internal and external spaces has maximised net usable floor area by reducing unnecessary circulation space, whilst at the same time allowing staff and patients to use the building in a controlled and efficient manner. The result is a lively working environment that is conducive to good health care and flexible to the changing needs of a clinical practice.

Sustainability and adaptive reuse at Norton Park, Edinburgh

Introduction

The selection and sensitive conversion of a former school to office and community centre has provided the opportunity for an imaginative project that exemplifies the balance that can be achieved between the conservation of a historic building and sustainability within the built environment (Fig. 6.14). Ongoing involvement by the design team and feedback into an information loop seeks to ensure the continued success and improvement of the project.

This case study is based on work undertaken by Burnett Pollock Associates of Edinburgh on behalf of the Albion Trust. Following completion in June 1998, the project was adopted by the Building Research Establishment (BRE) for its Energy Efficiency Best Practice Programme, and it was one of the UK's entries for the International Green Building Challenge 1998. In 1999 the conversion won the Regeneration of Scotland Award for demonstrating beneficial effects on the well-being and economy of the community, the Sir Robert Grieve Award as an outstanding example of sustainability, and a Civic Trust Award.

Selection and project objectives

Following selection and feasibility studies, Norton Park was chosen from a number of redundant schools in Edinburgh to provide accommodation for groups within the voluntary sector that had been identified as being poorly served by existing landlords. Ownership of the building was passed to the newly formed Albion Trust, and objectives for the project agreed – to achieve maximum sustainability in the wider sense of long-term social, economic and environmental factors; to provide flexible, healthy and accessible accommodation with shared facilities and resources; and to achieve a beneficial use for a historic building.

Social features

The project has helped to regenerate a deprived area of the city, whilst at the same time ensuring community involvement and interaction with local businesses and services. Consultants and contractors have been selected on the basis of both employment policies and practices, and attitudes towards community and environmental issues. Barrier-free access has been achieved throughout the building, with a lift and entrance ramps designed to provide maximise accessibility. A nursery and crèche are available for the occupants and wider community, and a resource centre offers information technology facilities.

Economic issues

Conversion of the school, rather than the erection of a new building, has been economically successful in terms of both capital and revenue costs, with design, construction and management decisions based on upstream and downstream costs for both finance and environmental impact. Life-cycle costings were calculated for new systems, materials and components,

and payback periods assessed for energy-conservation measures. Running costs have been minimised in terms of maintenance, energy and water-supply charges, and reductions in waste have reduced landfill tax liability.

Mezzanine floors, supported on glulam beams, have been used to increase the net lettable floor area from 1 940 m² to 2 542 m², offering shared facilities and office space for up to 30 charities of varying sizes.

Environmental benefits

Specified materials and components have been subjected to an environmental audit, with timber and timber products sourced from sustainable supplies; preservative treatments have not been used for either new or existing timbers. Plastics have been avoided wherever possible, and linoleum chosen in preference to vinyl for floor finishes. Use has also been made of natural-fibre carpeting, 'green' mineral paints and wax coatings, and an environmentally-acceptable compartment trunking system (Econet©) developed for power, telecommunication and data networks.

Losses to the original building fabric have been minimised, and obsolete fittings and components salvaged. Segregation of construction waste has also allowed for the recycling of waste materials and reuse of recovered items. Second-hand materials have been locally sourced wherever possible.

A rainwater recovery and retention system provides 'grey water' for toilet flushing, and recycling of future user wastes has been planned for by offering separated collection facilities. Soft landscaping is proposed as a replacement for existing asphalt surfaces, and gardens planned for additional learning, play and wildlife opportunities. Bat boxes are already in place to encourage species diversification.

Energy use

A significant upgrading of the total energy performance of the building has been achieved as compared with typical and good-practice energy consumption figures provided by the Department of the Environment, Transport and the Regions as part of the Best Practice programme (DETR, 1998) (Table 6.1). Mineral-wool insulation has been provided in the roof voids (300 mm), with eaves ventilation specifically designed for the stone-slate roof coverings, and to external walls (150 mm), giving U-values (thermal transmittance) of 0.12 W/m²K and 0.203 W/m²K respectively. Existing sash and casement windows have also been refurbished, and new secondary windows fitted with double-glazed argon-filled silver oxide-coated units, giving a U-value of 0.9 W/m²K.

Table 6.1 Annual delivered energy consumption (kWh/m² treated floor area) based on anticipated energy usage figures for Norton Park supplied by Ove Arup and Partners.

	Typical	Good practice	Norton Park
Heating and ventilation	240	130	78.42
Lighting (dependent on occupancy levels)	38	22	9.87

Heating is by a zoned low-pressure hot-water system, with condensing gas boilers and individual thermostatic valving. User-controlled direct fresh air ventilation in the summer and preheated ventilation, with heat recovery from extracted air and air drawn across the underside of the roofing slates, avoids the need for air conditioning. High-efficiency lighting, which is controlled by a building environment management system (BEMS), is responsive to daylight and occupancy levels. Adjustable solar shading shelves are also fitted.

Architectural features

The spaces and detailing within this listed (grade II) building have been sensitively handled, with traditional materials and repair practices used wherever possible (Fig. 6.15). Non-original features and inappropriate repairs have been removed, and new features, such as escape stairways, designed to harmonise with the building and its location.

Acknowledging the detrimental effects of previous repairs at Lincoln Cathedral

Introduction

The Special Repair Programme, undertaken between 1922 and 1932 in response to movement and instability in the medieval masonry structure of Lincoln Cathedral, has left a legacy of ill-conceived repair and change that continues to have a significant effect on current policy and practical conservation. The nature and extent of the work carried out under the project has recently been researched and shown to have been excessive and largely unjustified. Further investigation is recommended to establish the consequences of this work on the future well-being of the building.

Fig. 6.15 Norton Park interior. (Courtesy of Burnett Pollock Associates.)

This case study is based on research undertaken by Dr Michael O'Connor as part of his doctoral thesis entitled *Lincoln Cathedral: The Evolving Perception and Practice of Care in an Historic Masonry Structure* (O'Connor, 1998).

Background to the Special Repair Programme

The massive engineering campaign known as the Special Repair Programme, undertaken at Lincoln Cathedral between 1922 and 1932, was a response to concern over the stability of the medieval masonry structure (Fig. 6.16). Earlier works undertaken to strengthen the west front included the addition of two relieving arches in the eleventh century, the filling of interior stairways and passages in the eighteenth century, and the insertion of iron fixings in the early nineteenth century. These fixings, in the form of iron bands, have been calculated to add six tonnes to the weight of each of the two west towers, so accelerating the rate of deterioration and decay. In the latter part of the nineteenth century, flat wrought-iron wedges were driven into many of the masonry joints in order to correct the effects of earlier movement on the stonework. Corrosion and consequential expansion of these wedges have resulted in extensive structural distortion. Movement at this time was such that it was reported that 'tell-tales fixed one day were found to be broken in two the next'.

Fig. 6.16   West front, Lincoln Cathedral. (Lincoln Cathedral Works Archive.)

Nature and extent of the works

By the early years of the twentieth century, reports of structural weakness led to the adoption of a plan of work involving extensive pressure grouting with Portland cement and the insertion of concrete restraining beams within the western towers. In 1921 the process was described as 'a means by which cement is forced, like molten metal, under air pressure into the otherwise unreachable parts of the thickest walls, 20 or more feet in width ... rendering the whole structure monolithic'. Many thousands of

tons of Portland cement grout were introduced into the core of the masonry walls, encasing several miles of phosphor bronze reinforcing bar (see Fig. 3.32).

In a lecture given in 1931, the Special Repair Programme was explained as having 'forced into the walls and surrounds of the North West Tower 23 524 gallons of grout, equal to 3 764 cubic feet of space, and weighs approximately 107 tons, and into the Transepts and the Central Tower 12 682 gallons of grout'. In the following year, 'drilling of 23 miles of holes, all made with the impact of power-hammer drills, with more than a thousand tons of cement encasing 31 516 Delta cramps, and a total of 13 547 new stones had been fixed into the fabric' (Fig. 6.17).

Fig. 6.17   Drilling in advance of grouting as part of the Special Repair Programme (1922–32). (Lincoln Cathedral Works Archive.)

Consequences for the future

The effects of the Special Repair Programme on the medieval masonry structure of Lincoln Cathedral are only now being fully studied, and conclusions so far show the work to have caused irreversible damage (including direct damage to the Romanesque frieze) that was without justification. The consequences of this work on future cycles of repair and maintenance have not yet been fully realised, although further investigation into the nature and extent of the work is considered essential.

In the first instance it has been recommended that the extent of the Special Repair Programme be further researched and recorded to provide a comprehensive analysis of resulting structural implications. This would require a combination of archaeological survey and conservation condition report, with strategic excavation undertaken to confirm the presence of reinforcement and grouting.

Long-term monitoring, using a three-dimensional grid-plan of the building, would provide a comprehensive visual record of the constantly changing state of the structure, including past and future interventions and movements. Such a record would also allow normal cyclical movements, including those resulting from seasonal changes, to be established so that anomalous movements might be identified. The resulting three-dimensional record would then be used to inform structural analysis, and the specification and implementation of appropriate remedial action.

Bringing a ruin back to life at Houghton-on-the-Hill, Norfolk

Introduction

The ruinous and tumbling remains of earlier buildings can evoke strong feelings and provide an opportunity to understand a little of the lives of those who built and lived in them. The consolidation and repair of St Mary's Church, which has extended the life and utility of an important medieval building, has contributed to present knowledge through the study of its construction and remaining wall-painting schemes.

This case study is based on work initially undertaken by Norfolk County Council as part of its *Ruined Churches Repair Programme*, which resulted in the consolidation and repair of eleven churches (Watt, 1996, 1997), and subsequent phases of work undertaken on behalf of the Friends of St Mary's. The project at Houghton-on-the-Hill was awarded the joint Royal Institution of Chartered Surveyors' (RICS) Building Conservation Award for 1998.

Abandonment and dereliction

Norfolk possesses an astonishing number of medieval parish churches, with at least 921 being built between the eleventh and sixteenth centuries. Of these churches, 610 (66.2%) are still in regular use; 59 (6.4%) are fully standing, but no longer used; 100 (10.9%) are in ruins; and 152 (16.5%) have no above-ground remains.

St Mary's Church, which stands isolated in a rural wooded setting, was built in the late eleventh century, with the west tower added in the fifteenth century and the present chancel dating from the eighteenth century. By 1760 the village had almost ceased to exist, and a survey carried out in 1949 revealed the church to have recently fallen out of use. By 1953 it was reported that 'holes in the roof and windows let in the rain and the wind, ivy and brambles surge over the walls and unhampered through the windows. Plaster is crumbling from the walls to form a damp and dirty carpet'.

Ruined Churches Repair Programme

The works undertaken to selected churches within the Ruined Churches Repair Programme followed a general phased approach over a three-year period, with site clearance, initial recording and emergency works in phase one; full consolidation and repair in phase two; and interpretation, additional recording and archaeological investigation in phase three.

At the start of the first phase, the church was almost totally covered with live ivy growth (Fig. 6.18), denying both inspection and access. Vandalism was evident and local rumours of Satanic activities were commonplace. The ivy was cut and trimmed to allow entry, and the main stems severed to allow the growth to die back naturally. Much of the tile roof covering to the nave was missing and three tie beams were severely decayed. The slated chancel roof covering was similarly defective, and the exposed roof carcass, together with an earlier and lower roof, was close to collapse.

Emergency works were undertaken during 1993, consisting of introducing additional timbers to tie the feet of the principal rafters and reduce the risk of spread, securing roofing felt over the original ties to lessen further deterioration, and installing temporary supports beneath the chancel roof and to two niches in the east nave wall.

Phase two works undertaken during 1995–6 included the removal of dead vegetation growth, resetting and pointing of flints and stones, repointing open joints and replacing defective mortar, rough-racking

Fig. 6.18   Ivy growth in advance of phase one clearance.

broken wall faces and wall heads, structural repairs, repairing and re-
covering the nave and chancel roofs, building up fallen areas of flint facing,
building up honeycomb brick blockings to window openings, and limited
site clearance.

Preliminary investigation of partially visible wall paintings and a later
survey of wall surfaces revealed large areas of early medieval figura-
tive wall painting considered contemporary with the construction of the
church towards the end of the eleventh century (Fig. 6.19). Photogram-
metric recording, environmental monitoring, liquid moisture survey and
emergency works were undertaken to document and ensure the stability
of the painted surfaces.

With the recognised importance of the wall paintings and the need to
stabilise the internal environment in advance of long-term monitoring,

Fig. 6.19   Wall painting to the east nave wall.

additional works were undertaken during 1997–8 that comprised the formation of a new roof to the tower, insertion of two new floors within the tower, fitting of window louvres and mesh screens to the tower windows, provision of a lightning protection system and fitting of a new entrance door (Fig. 6.20).

Phase three works were expanded to allow for a programme of detailed investigation and recording undertaken between 1997 and 2001 to inform subsequent works. This included additional environmental monitoring, archaeological evaluation, topographical survey and technical reports (including biodeterioration, soluble salts and pigment analysis). Conservation and management plans were adopted in 2001 and 2002 respectively, which provided the basis for specific interventions including the installation of a rainwater disposal system, reinstatement and glazing of window frames and tracery, and installation of new flooring to chancel, nave and tower.

The final phase of works commenced in 2006, which included the conservation and reintegration of wall paintings and plaster layers, installation of electrical power and lighting, provision of new seating, and improved interpretation and presentation.

Fig. 6.20   Completion of phase two works.

Giving new life to an old building

Given the condition of St Mary's Church at the start of the initial programme, and with experience gained from other similar projects, it is clear that without intervention the church would have deteriorated and suffered structural failure. St Mary's Church has been given a new life and, with the wall paintings, is attracting a wide range of visitors who are keen to see the work that has been carried out. Responsibility for the church has been taken up by the Friends of St Mary's, and religious services are again being held on the site. The church has been brought back into beneficial use, which has generated a strong sense of ownership in a rural community.

Managing change within the Willis Corroon Building, Ipswich

Introduction

The management of change affecting historic buildings is a complex issue that has to recognise the special architectural and historic interest of the building within the framework of current listed building legislation, whilst at the same time respecting the original design philosophy and responding to the day-to-day needs of the owner. The guidelines that have been developed to manage changes to the award-winning Willis Corroon building in Ipswich have ensured that the building continues to be used, whilst ensuring that its special character is protected from potentially damaging alterations and change.

This case study draws on information published by Bob Kindred in *Context* (1995) and *Modern Matters: Principles and Practice in Conserving Recent Architecture* (1996).

Willis Corroon Building

The Willis Corroon building was designed by Norman Foster for the international insurance brokers Willis Faber and Dumas (now known as Willis plc) in 1974–5. The design of the building included a swimming pool, snack bar and gymnasium for the employees on the ground floor, open-plan offices using partitioning, and a roof-top restaurant and garden, and employed curtain wall glazing to create an imaginative working environment and bold architectural statement (Fig. 6.21).

Responding to the threat of a proposal to increase office space by expanding into the then little-used swimming pool area, the building became the subject of an intense lobbying campaign by local and national groups and was spot-listed as a building of special architectural or historic interest (grade I) in 1991.

Although internal alterations had taken place since completion in 1975, much had been within the spirit of the original design. Future changes would, however, be subject to the scrutiny of the local authority and English Heritage, and planning permission and/or listed building consent required where appropriate.

As a business in the international insurance market, Willis were concerned about the timescales needed for listed building consent to undertake the kind of internal replanning sometimes necessary to respond to a major claim. A set of agreed guidelines would be a way of managing change outside the statutory process, thereby reducing the company's uncertainty.

Fig. 6.21 Willis Corroon building at night, showing use of curtain wall glazing and internal demountable partitioning. (Photograph by Bob Kindred.)

Guidelines for change

It was as an aid to defining what alterations would or would not require planning permission and/or listed building consent that guidelines were first drawn up and adopted by all parties in February 1992. These placed particular emphasis on considering future proposals in the context of the original design philosophy, the framing of listed building legislation and the business needs of the owner.

The guidelines provide a statement of intention, with objectives and general principles worded in such a way as to respect and respond to the changing needs of the building owner. Clear guidance is given on the use of the various parts of the building, and on specific items including partitions, interior perimeter glazing, floor coverings, lighting and ceiling coverings, exterior glazing (Fig. 6.22), roofing and the curtilage area.

A statement is attached to each section, which defines the changes that will or will not require listed building consent. The building is essentially open-plan on the first and second floors, but partly divided into cellular offices on the ground floor by a demountable, full-height partitioning system. Here changes to the overall envelope would require consent, but rearrangement and/or subdivision using the partitioning system would not require consent. In the open-plan offices, no office furniture would

Fig. 6.22   Changes to the pattern and type of glazing are part of the management guidelines. (Photograph by Bob Kindred.)

be more than two-thirds the height of the space. Where changes are being considered to the distinctive internal finishes or colour scheme, early discussion with the local authority is required.

Effective management of change

The management of change within a modern working building, which is also listed for its special architectural or historic interest, has been achieved by practical 'management by agreement'. This allows changes to be defined in terms of their effect on the special qualities of the building, whilst being aware of the everyday needs of a business activity.

Guidelines have to be flexible and responsive to the needs of a particular building and its owner. As such, they may include some or all of the following:

- status of the document (i.e. who prepared it and who has agreed to it)
- definition of items of special interest
- analysis of the design philosophy
- details of context, plan form, structure, materials, external and internal treatments, decorative schemes and finishes, fixed integral plant and machinery, and fixed original works of art

- future intentions in programmes of demolition, alteration and extension
- future intentions in modification or replacement of plant and machinery
- legal commentary
- definition of works requiring require listed building consent
- definition of works requiring planning permission
- delineation of curtilage
- details of associated buildings, structures or landscape features
- timetable for review of document

The methodology embodied in the Willis guidelines was seen by government as having the potential to simplify controls on historic building alterations and a form of positive, proactive management of heritage resources. In 2003, English Heritage and the Office of the Deputy Prime Minister commissioned Paul Drury to investigate the effectiveness of management agreements (with the Willis building forming one of the three case studies). It was concluded that, although such mechanisms were effective for the few buildings to which they had been applied, they were essentially advisory (Drury *et al.*, 2003). Primary legislation would be required for them to have statutory force, and these now form part of the proposed reforms of heritage legislation under the Heritage Protection Review intended to come into force in 2010.

The PHAROS project and Happisburgh Lighthouse

Introduction

The PHAROS lighthouse project, funded by the European Commission Culture 2000 programme during 2004–7, is intended to raise awareness for historic lighthouses and the role they play as part of the coastal and marine heritage. Project objectives include understanding generic and specific causes of deterioration and decay, providing guidance on appropriate methods of repair and conservation, and arranging exhibitions and conferences to raise awareness amongst younger generations in the partner countries of Greece, Cyprus, Italy, Norway and the UK.

The lighthouse chosen as the UK case study is at Happisburgh in Norfolk. Selection was based on its early date relative to those of the other partner countries, its independent status with operation as a navigational aid through the Happisburgh Lighthouse Trust, and its existing interpretation and presentation to the public through the activities of the Friends of Happisburgh Lighthouse.

Happisburgh lighthouse

Happisburgh lighthouse is one of two lighthouses constructed following a severe storm in October 1789. The lighthouses came into operation in January 1791, with each lantern carrying numerous lighted candles so that by keeping the lights in line with each other, ships could pass the southern tip of Haisborough Sands and enter sheltered water between the Sands and the shore.

When Robert Stevenson visited Happisburgh in 1801, the candles had been replaced by Argand oil lamps with polished reflectors. A revolutionary helically framed lantern was installed in 1863, and cannel gas lighting replaced the Argand lamps in 1865. Not being sufficiently close to a mains supply, gas was manufactured within the grounds of the high light using five coal-fired retorts and two small gasholders. A new optic was fitted in 1868, giving the high light a range of 27 km compared with a range of 24 km for the low light. The fixed beacon of the high light was changed to an occulting character in 1883 and, with an extra light-vessel guarding Haisborough Sands and a severe threat of cliff erosion, the low light was discontinued in 1883 and subsequently demolished. Cannel gas lighting was replaced with an incandescent paraffin-vapour burner in 1910 and, with Happisburgh lighthouse being changed to an 'unwatched' beacon in 1929, the use of acetylene as the light source made it possible to dispense with the need for resident keepers. Two adjoining single-storey keepers' cottages were sold and became private dwellings. Electricity was installed in 1947, and the acetylene system retained as a secondary standby light.

In a review of navigational aids during 1987, the Trinity House Lighthouse Service, as the General Lighthouse Authority for England, Wales, the Channel Islands, and Gibraltar, announced its intention to close Happisburgh lighthouse. A local campaign won an initial extension and, with the necessary parliamentary bill to create an Established Lighthouse Authority, the lighthouse was transferred to the Happisburgh Lighthouse Trust in August 1990. It is the oldest working lighthouse in East Anglia, and the only independently run operational lighthouse in the UK (Fig. 6.23).

Understanding construction and condition

An accurate dimensional survey of the lighthouse provided metric data for the construction of a three-dimensional computer model of the structure, allowing digital visualisation and animation that sets the lighthouse in the context of its coastal site (Fig. 6.24).

A preliminary condition survey of the lighthouse raised a series of questions, including confirmation of construction beneath external and internal render layers, changes in construction over the height of the structure, the

Fig. 6.23 Happisburgh lighthouse.

Fig. 6.24   Computer visualisation of lighthouse. (Image by UKgeomatics/Arcvisuals.)

presence/absence of metallic cramps or other reinforcement, and penetration of stone cantilevered steps and structural members. It was also necessary to know whether there were voids within the wall thickness, and if there was separation of the external and internal render layers from the substrate. Non-destructive survey of sample areas of the lighthouse was undertaken in 2006, using a combination of impulse radar, dynamic impedance, thermography, and pulsed ultrasound, with selective opening-up to confirm survey data. These investigations have confirmed that the walls of the lighthouse are of brick, with cement-based external and internal render layers. There are no metallic cramps or reinforcement, but there is evidence of voids within the brickwork and separation of the render from the substrate. The stone treads penetrate the wall by c. 390 mm (overall wall thickness 1.1 m) and display soluble salt damage, whilst the composite cross-members beneath the service and lantern rooms penetrate the wall by 400–450 mm (overall wall thickness 0.9 m).

Environmental monitoring is being undertaken within the lighthouse to help understand how the structure responds to external conditions and to inform future decisions regarding heating within the lighthouse.

Hydrocarbon spillage and its effect on stone monuments

Introduction

Environmental pollution has a significant effect on buildings and monuments, with much research having been undertaken to consider resulting decay mechanisms affecting porous materials. The effects of pollution on below-ground fabric has, by contrast, been less widely considered and, in

the case of chronic petroleum-hydrocarbon leakage from refineries situated close to stone monuments, little is known about the resulting processes and methods of remediation.

This case study is based on research undertaken by Dr Maria Kliafa, under the supervision of the author, as part of her doctoral thesis entitled *An Investigation of Bioremediation for the Conservation of Petroleum-Contaminated Stone Monuments* (Kliafa *et al.*, 2003; Kliafa, 2005).

Hydrocarbon contamination

Groundwater is associated with various decay mechanisms – including rising damp, soluble salt mobilisation, freeze-thaw cycling, ground heave, and flooding – and can act as the carrier of various pollutants such as pesticides, agricultural and industrial waste products, and hydrocarbons. The treatment of contaminated soil and groundwater can have significant and costly implications for developments on new and brownfield sites, but for existing buildings and structures the effects on durability and performance over time cannot readily be determined.

Hydrocarbon pollution may be attributed to leakage, spillage and accidental discharge from liquid fuel storage and distribution facilities; extraction and storage of solid fuels; improper storage and use of industrial solvents; and uncontrolled disposal of used lubrication oils. Spillage from the *Exxon Valdez* in 1989 resulted in 10.8 million US gallons of crude oil being discharged into Prince William Sound and affected 1900 km of Alaskan coastline; whilst the grounding of the *MV Braer* on Garth's Ness in the Shetland Islands in 1993 led to the discharge of 85 000 tonnes of crude oil and 1500 tonnes of bunker (heavy fuel) oil. It is reported that 500 000–1 700 000 tonnes of hydrocarbons enter the Mediterranean sea each year, and that the Niger Delta is one of the five most polluted sites on the planet, with up to 1.5 million tones of oil having entered the delta over the past 50 years.

Research undertaken in the 1970s showed that the soil and groundwater in the archaeologically important area of Eleusina in Greece was contaminated by petroleum hydrocarbons as a result of leakage from two petrol refineries and inflow from polluted seawater. Further refinery leakage was detected in the 1980s and concern expressed for the nearby Lake Koumoundourou, which was characterised as an area of significant archaeological interest and natural beauty.

Remediation techniques

Whilst techniques for the treatment of hydrocarbon-polluted soils are well established, there have been no published data to determine

their effectiveness in relation to the subterranean elements of masonry monuments. In addition, where these monuments are deemed to be of archaeological or historical significance, techniques of remediation must respect the inherent values of the fabric and site.

The treatment of polluted soils may be considered in terms of *ex situ* or *in situ* methods. *Ex situ* methods require the soil to be excavated or groundwater extracted, pollution reduction measures based on biological or non-biological treatments undertaken either on or off site, and the solid material returned. Such methods allow for process control and monitoring, and are comparably quicker to perform. *In situ* methods are based on physical/chemical processes (i.e. air sparging, hydraulic and pneumatic fracturing, washing), electrokinetics, or biotreatment (i.e. bioslurping, phytoremediation, bioremediation).

Bioremediation

Research to determine the most appropriate technique for the treatment of hydrocarbon-contaminated subterranean fabric considered both the applicability of the methods and their impact on the archaeological and historical significant of the monument.

In situ bioremediation, which makes use of biological micro-organisms to degrade organic contaminants, was pioneered in the early 1970s in response to petroleum spillage. The micro-organisms gain energy by catalysing energy-producing chemical reactions, which involve breaking chemical bonds and transferring electrons away from the contaminant in an oxidation-reduction reaction. Of the various bioremediation methods, bioventing was considered the most appropriate for this particular research programme on the basis of its *in situ* application, proven use in the treatment of petroleum-hydrocarbon ground contamination, compatibility with natural stone masonry, and environmental safety.

Experimental bioventing

Bioventing is a method that aerates the soil to stimulate *in situ* biological activity and to promote bioremediation. Air is typically injected at a low rate through a well into the contaminated soil, with monitoring wells allowing for analysis of off-gases and determination of rates of remediation. There are many variables, including soil gas permeability, contaminant distribution, microbial activity, soil characteristics, and the biodegradability of the targeted organic compounds.

Laboratory tests applying bioventing to samples of three hydrogen-impregnated limestones selected for their particular physical and mechanical characteristics as compared with the subterranean fabric of

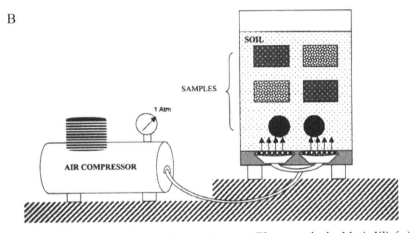

Fig. 6.25 (a,b) Laboratory bioventing equipment. (Photography by Maria Kliafa.)

stone monuments indicated that the method has potential for the *in situ* treatment of petroleum-hydrocarbon contaminated subterranean stone masonry (Fig. 6.25). Results showed that bioventing was 156 times more efficient at removing petroleum-hydrocarbons from buried stone samples than reliance on natural decomposition.

The history and performance of gauged brickwork

Introduction

Gauged brickwork – the term used in England since the seventeenth century to describe brickwork of a 'superior finish required to provide detail

to important elevations, including moulded reveals, arches, string courses and other forms of ornamentation' – developed out of the earlier cutting and rubbing of fired bricks to achieve a precision and fineness of jointing that had its origins in Flanders of the fourteenth century. The practical skills of gauged brickwork reached their peak in England during the second half of the seventeenth century and again in the latter part of the nineteenth century, alongside the revival of earlier architectural styles and associated embellishment.

De-skilling within the construction industry over the past 25 years has seen a steady decline in the craft skills required to produce gauged brick-work, together with the production of suitably soft rubbing bricks, and much that was once commonplace has become almost lost to the construc-tion industry. The repair and conservation of existing buildings, together with sensitive new development in historic settings, demand skills and materials that deserve to be better understood.

This case study is based on research undertaken by Dr Gerard Lynch, under the supervision of the author, as part of his doctoral thesis entitled *English Gauged Brickwork: Historical Development and Future Practices* (Lynch, 2004; Lynch *et al.*, 2006).

Brick production and use

The rubbing bricks or 'rubbers' used in gauged brickwork were made with finely graded and naturally washed alluvial brickearth and clays having a typically high silica content (>80%) that was readily available over large parts of southern England. This material would have been tempered with sand, moulded and clamp- or kiln-fired using wood as a fuel at a tem-perature of 850–950°C. This temperature is significant, as at about 900°C vitrification begins and a fireskin develops on the brick. Below this tem-perature a brick may best be described as 'baked', whilst modern firing temperatures (>1000°C) are too high to produce rubbing bricks.

A small number of traditional brickmakers continue to supply rubbing bricks, but limited use of naturally high silica-bearing brickearths and re-liance on coal and liquid petroleum gas-fired kilns with temperatures in excess of 1000°C mean that availability of good-quality bricks for gauged work is limited, and costs are correspondingly high.

Rubbing bricks would originally have been rubbed square on their bed and face on a flat rubbing stone, and then shaped to a template using a variety of tools including double-bladed brick axes, brick scotch, small hand saws, and a variety of files and abrasives. Later, in the late nine-teenth century, the bricks would instead have been secured in a profiled timber cutting box and cut to shape using a bow-saw with a twisted wire

Fig. 6.26 Gauged brick niche constructed by Gerard Lynch in the manner of an apprentice 'masterpiece' as part of his doctoral programme. This scaled piece was constructed using six smaller bricks that were individually cut and shaped from one original oversized TLB orange rubbing brick that was produced in the 1950s.

blade before being finished using abrasives. The precisely shaped rubbing bricks would be soaked in clean water and set by 'dip-laying' on to a thin lime-putty and silver-sand mortar to form a fine joint of 1–2 mm width (Fig. 6.26).

Durability and performance

Despite being fired to a point below vitrification and being without a protective fireskin, soft rubbing bricks have clearly been durable enough to withstand site conditions over hundreds of years. Research to determine the performance of modern rubbing bricks compared with that of their historic counterparts was undertaken to inform current uses and future production.

Samples of traditional English, Dutch and Belgian rubbing bricks from the seventeenth to twentieth centuries were subjected to mineralogical and

petrographic analyses, with the results showing clear physical differences that, in part, relate to firing temperatures.

The durability of traditional rubbing bricks appears to relate primarily to their high porosity associated with a low firing temperature. Furthermore, the greater ability of historic rubbing bricks to absorb water by capillary action implies that their fine pores are better interconnected and more effective at transporting water than modern bricks.

Climate change and the historic environment

Introduction

Climate is a key factor in both design and occupation, and has prompted some of the most innovative forms of construction in the history of mankind. One need only observe the extreme conditions faced by arctic or desert communities to see how an understanding of climate has served to ensure levels of protection and comfort without which continued habitation in some of the world's most hostile environments would not be possible.

The effects of climate change, and what impact this will have in the short, medium and long term on both natural and man-made environments, is perhaps one of the most pressing issues of the present generation. Without fully understanding the underlying and contributory causes, and developing a realistic framework for assessing the effects on what we consider important, then change will be forced on how we choose to live and work.

This case study draws on research published by the Centre for Sustainable Heritage, University College London, in *Climate Change and the Historic Environment* (Cassar, 2005), which was commissioned by English Heritage in 2002 as a scoping study to begin mapping the effects of climate change.

The changing climate

The Intergovernmental Panel of Climate Change (IPCC) states that 'An increasing body of observations gives a collective picture of a warming world and other changes in the climate system' and 'emissions of greenhouse gases and aerosols due to human activities continue to alter the atmosphere in ways that are expected to affect the climate'. The global average surface temperature increased over the twentieth century by about 0.6°C; temperatures have risen during the past four decades in the lowest 8 km of the atmosphere; snow cover and ice extent have

decreased; global average sea levels have risen and ocean heat content has increased; and changes have occurred in relation to the frequency and intensity of precipitation, drought and flooding, and temperature extremes.

In the UK, the UK Climate Impacts Programme (UKCIP) makes it clear that expected climate change over the next 30–40 years is largely the result of past greenhouse gas emissions, with subsequent changes determined by current emissions of carbon dioxide and methane. The message is that we need to adapt our ways of living to prepare for changes that are already in the climate system and limit future emissions of greenhouse gases and aerosols.

As the historic environment typically operates on long-term planning, and for which climate change has particularly serious implications, heritage organisations have been active in assessing the impacts of climate change in relation to future management and planning. The National Trust, for instance, with its varied portfolio of natural and manmade assets, is undertaking studies in relation to soils, flora, fauna, landscape, agriculture, gardens, forestry, water resources, energy, buildings, health, recreation, heritage, and coastal zones, and has established policies to protect its resources.

Key climate factors

Research conducted by the Centre for Sustainable Heritage to determine heritage susceptibility to climate change focused on three groups: buildings (i.e. all built and exposed heritage, including excavated archaeological sites), archaeology (i.e. buried archaeological sites, including potential sites), and parks and gardens in two different English regions – the north east and the east of England.

The climate change factors of greatest concern for the historic environment were considered to be:

- *temperature* – clear outcomes, with likely increases in deterioration mechanisms and much being dependent on the rate of change
- *reduced spring/summer/autumn rainfall* – problems stemming from agricultural competition for limited resources, especially in the east of England
- *extreme rainfall in winter and high winds* – extreme winds and storminess might not be predictable, but have enormous significance in future planning
- *fluvial and runoff flooding* – widespread effect, with responses likely to impact on historic sites

Fig. 6.27   Significant coastal erosion with unsustainable sea defences and preferred policy of unabated erosion leading to loss of heritage assets.

- *coastal loss and flooding* – potential devastation of vulnerable sites, with damage likely to result from adaptation measures to protect other parts of the coastline (Fig. 6.27)

Key recommendations

In relation to the fragile heritage that makes such an important contribution to the rural and urban living, eight recommendations were put forward for consideration:

- *Sector leadership on climate change* – English Heritage should maintain sector leadership on climate change through developing climate change indicators, disseminating practical information to historic environment stakeholders, and promoting the inclusion of climate change impact in wider agendas.
- *Monitoring, management and maintenance* – current monitoring, management and maintenance practices should be revised to improve the stability of the historic environment. This includes promotion and support for local decision-making in maintenance and emergency response, with cross-disciplinary training programmes and an emphasis on grants for maintenance rather than repair.

- *Value and significance in managing climate change impacts* – value and significance must be considered in the future planning of the historic environment, as it is not realistic to conserve anything for ever or everything for any time at all.
- *Participation in the planning of other agencies* – English Heritage should participate and contribute in relation to measures being developed by other agencies responding to climate change impacts in other sectors.
- *Fully functional heritage information system* – a fully integrated heritage information system should replace individual paper maps and disparate databases, with the ability to capture, display and analyse data in context with other geographic data and information from climate change models. Such a system should be capable of continuous upgrading and refinement and be available as an on-line facility.
- *Emergency procedures* – a coordinated damage alleviation service should be available to deal with extreme weather affecting the historic environment, as is available in other countries. A key function of such a service would be to provide immediate protection to storm-damaged property to reduce the risk of further damage.
- *Adaptation strategies and guidelines for historic buildings, archaeology, parks and gardens* – guidance should be made available on how current conservation and maintenance practices should respond to climate change. This should include dissemination and integration of research into existing or planned projects, including the adaptation of drainage and rainwater goods, together with the provision of irrigation and water storage.
- *Buried archaeology and prediction maps* – prediction maps should be developed to draw on a wide number of interrelated variables, rather than on a single variable such as oil type.

Lessons to be learned

The case studies presented in this chapter demonstrate how buildings respond to the various agencies or mechanisms of deterioration and decay, and how solutions to these and other issues have to be based on a detailed understanding of the buildings and the people who use them.

Above all, it is important to acknowledge that there is no such thing as a standard solution to the problems that affect the performance and use of buildings. This is particularly true when dealing with those of architectural or historic importance. Every building has therefore to be considered on its individual merits, based on the knowledge gained through appropriate

levels of survey and assessment. The management and aftercare of buildings, which seeks to ensure best use and practice, will be covered in Chapter 7.

References

Blackney, K. & Martin, B. (1998) The application of cathodic protection to historic buildings: Buried metal cramp conservation in the Inigo Jones Gateway, Chiswick House Grounds, London. *English Heritage Research Transactions – Volume 1: Metals*. London: James & James.

Building Research Establishment (1981) *Rising Damp in Walls: Diagnosis and Treatment*. Digest 245. Garston: BRE.

Burland, J.B. (2001) The stabilisation of the leaning tower of Pisa. *Ingeni*, **10**, 10–18.

Cassar, M. (2005) *Climate Change and the Historic Environment*. London: Centre for Sustainable Heritage, University College London.

Colston, B., Watt, D. & Munro, H. (2001) Environmentally-induced stone decay: the cumulative effects of crystallisation-hydration cycles on a lincolnshire oopelsparite limestone. *Journal of Cultural Heritage*, **4**, 297–307.

Cope, B., Garrington, N., Matthews, A. & Watt, D. (1995) Biocide residues as a hazard in historic buildings: Pentachlorophenol at Melton Constable Hall. *Journal of Architectural Conservation*, **1** (2), 36–44.

DETR (1998) *Energy Use in Offices*. Energy Consumption Guide 19. London: DETR.

Drury, P., McPherson, A. & Allen, R. (2003) *Streamlining Listed Building Consent: Lessons from the Use of Management Agreements. A Research Report*. London: English Heritage/ODPM.

Kindred, B. (1995) Pioneering management guidelines for modern listed buildings: The Willis Corroon Building, Ipswich. *Context*, **47**, 12–15.

Kindred, B. (1996) Management issues and willis corroon. In *Modern Matters: Principles and Practice in Conserving Recent Architecture* (S. Macdonald ed.). Shaftesbury: Donhead Publishing.

Kliafa, M., Colston, B. & Watt, D. (2003) Evaluating the use of bioremediation techniques in the conservation of hydrocarbon-contaminated stone monuments. In *Conservation Science 2002* (J.H. Townsend ed.), pp. 135–40. Archetype Books, London.

Kliafa, M. (2005) *An Investigation of Bioremediation for the Conservation of Petroleum-Contaminated Stone Monuments*. Unpublished PhD thesis. Leicester: Faculty of Art and Design, De Montfort University.

Lowe, L. (1997) *Building in Partnership: Earthquake Resistant Housing in Peru*. Rugby: Intermediate Technology.

Lynch, G. (2004) *English Gauged Brickwork: Historical Development and Future Practices*. Unpublished PhD thesis. Leicester: Faculty of Art and Design, De Montfort University.

Lynch, G., Watt, D. & Colston, B. (2006) An investigation of hand tools used for english cut-and-rubbed and gauged brickwork. In *Proceedings of the Second International Congress of Construction History* (M. Dunkeld, J. Campbell, H. Louw, M. Tutton, B. Addis & R. Thorne eds.), pp. 2017–36. London: Construction History Society.

O'Connor, M. (1998) *Lincoln Cathedral: The Evolving Perception and Practice of Care in an Historic Masonry Structure*. Unpublished PhD thesis. Leicester: School of the Built Environment, De Montfort University.

Watt, D.S. (1996) Consolidation and repair of standing ruins: Medieval Churches in Norfolk – Parts I and II. *Structural Survey*, **14** (3/4), 10–16/48–55.

Watt, D.S. (1997) The consolidation and repair of St Mary's Church, Houghton-on-the-Hill, Norfolk. *Transactions of the Association for Studies in the Conservation of Historic Buildings*, **22**, 31–39.

Watt, D. & Colston, B. (2000) Investigating the effects of humidity and salt crystallisation on medieval masonry. *Building and Environment*, **35**, 737–49.

Chapter 7
Building Management and Aftercare

Planning the future

Referring back to the definitions of building pathology given at the start of Chapter 1, it is apparent that the design and implementation of remedial works, together with monitoring and evaluation, form an integral part of the overall study and practice of building pathology.

Whilst it is not possible to consider these tasks in isolation from the circumstances of a particular building or project, it is the intention of this final chapter to identify some of the key issues that will affect the use and well-being of buildings and their occupants as a result of such action.

What can be done with buildings?

Buildings are used for a vast range of activities, and each must satisfy its owner or user in order for it to achieve social acceptance or commercial viability (Fig. 7.1). Some, such as historic buildings or monuments, may survive on the basis of protective legislation or inherent charm, whilst others have to provide facilities and space to meet the demands of a fickle population and variable workforce.

In order for a building to be matched to the needs of a current owner or user, it is inevitable that, during the course of its life, it will be subject to some form of change. This might take the form of a simple extension to satisfy the needs of a growing family or involve complex alterations to structure, fabric and services in order to create suitable conditions for a new or updated use. How these changes are planned, managed and implemented will be crucial to the eventual success of the building, and will thus require careful consideration of both the building and its intended owner or user.

Matching the needs of a client or tenant with what a building can realistically provide is both challenging and rewarding, and the processes of

Fig. 7.1   Perlan (The Pearl), Öskjuhlio, Reykjavík, Iceland. Hot-water storage tanks for the district heating system form the basis for a winter garden, restaurant, meeting room and viewing platform. The dome consists of reflective glass panels on a hollow steel frame carrying hot or cold water to regulate the temperature inside.

briefing and selection need to be fully understood. In this, it is inevitable that people have become influenced by modern buildings, and the conveniences and facilities that are offered, and guidance is needed in the choices and levels of expectation. This should include:

- determination of user requirements (e.g. technical, functional, behavioural)
- identification of limitations of chosen building (e.g. structural, spatial, locational, environmental)
- matching activity to building (i.e. not vice versa)
- working with the building and not against it (i.e. be sympathetic to existing spaces and restrictions)

The nature and extent of building works will, to a greater or lesser extent, reflect the needs of the building and/or its owner or user, and thus vary in how much it changes the building and its facilities. Works may thus be considered as attempting to retain the *status quo* through activities such as:

- *Conservation* – making a building fit for 'some socially useful purpose' (International Council on Monuments and Sites, 1964); '. . . action taken to prevent decay and manage change . . . embraces all acts that prolong the life of our cultural and natural heritage' (Feilden, 2003, p. 3); 'action to secure the survival or preservation of buildings, cultural artefacts, natural resources, energy, or any other thing of acknowledged value for the future' (BS 7913:1998); 'process of managing change in ways that will best sustain the values of a place in its contexts, and which recognises opportunities to reveal and reinforce those values' (English Heritage, 2006).
- *Preservation* – 'method involving the retention of the building or monument in a sound static condition, without any material addition thereto or subtraction therefrom, so that it can be handed down to futurity with all the evidences of its character and age unimpaired' (Baines, 1923); 'to keep safe from harm' (English Heritage, 2006).
- *Repair* – 'restoration of an item to an acceptable condition by the renewal, replacement or mending of worn, damaged or decayed parts' (BS 8210:1986); 'work beyond the scope of regular maintenance to remedy defects, significant decay or damage caused deliberately or by accident, neglect, normal weathering or wear and tear, the object of which is to return the building or artefact to good order' (BS 7913:1998).
- *Maintenance* – 'combination of all technical and administrative actions, including supervision actions, intended to retain an item in, or restore it to, a state in which it can perform a required function' (BS 3811:1993); 'routine work necessary to keep the fabric of a building, the moving parts of machinery, grounds, gardens or any other artefact, in good order' (BS 7913:1998).
- *Reconstruction* – reassembling a building using 'extant materials and components supplemented by new materials of a similar type, using techniques approximating to those believed to have been used originally, based on existing foundations and residual structure, historical or archaeological evidence' (British Standards Institution, n.d.)

Building works may also seek to change the building and/or its facilities through one or more of the following processes:

- *adaptation* – accommodating a change in the use of a building, which can include alterations and extensions

- *alteration* – changing or improving the function of a building to meet new requirements
- *conversion* – making a building of one particular type fit for the purposes of another type of usage
- *extension* – increasing the floor area of a building, whether vertically by increasing height or horizontally by increasing plan area
- *improvement* – bringing a building and/or its facilities up to an acceptable standard, possibly including alterations, extensions or some degree of adaptation
- *modernisation* – bringing a building up to a standard laid down by society and/or statutory requirements
- *refurbishment* – overhauling a building and bringing it up to current acceptable functional conditons
- *rehabilitation* – work beyond the scope of planned maintenance, to extend the life of a building, which is socially desirable and economically viable
- *relocation* – dismantling and re-erecting a building at a different site
- *renovation* – restoring a building to an acceptable condition, which may include works of conversion
- *restoration* – restoring the physical and/or decorative condition of a building to that of a particular date or event
- *revitalisation* – extending the life of a building by providing or improving facilities, which may include works of repair

Managing building and change

Going back to the idea that buildings are formed as a series of layers that are affected by different rates of change (see Chapter 2), those proposed by Duffy (1990, p. 17) are allocated their own timespans – *shells* last up to 50 years, *services* last up to 15 years, *scenery* lasts up to 5 years, and *sets* change on a daily basis. Awareness and control of such change relies on appropriate management techniques – whether of buildings, people or their activities – and the development of matching skills and disciplines.

Of all the management techniques – cost-in-use calculation, cost-benefit analysis, whole life-cycle cost analysis (Table 7.1), information management, maintenance management – available to those responsible for the use and aftercare of buildings, it is perhaps facilities management that has the most general applicability. Facilities management is a broad-based management approach to looking after buildings and people, and may be defined as 'the practice of coordinating the physical workplace with people and work of the organisation ... integrates the principles of business

Table 7.1 Comparative life-cycle costs for roofing products (Macdonald *et al.*, 2003, p. 17).

Product	Durability (years)	Initialcost (£) for supply and fixing/m²	Repair and maintenance (% of initial cost)	Cost in 100 years cycle (£/m²)
Aluminium sheet	40	32.50	2	102.00
Copper sheet	65	42.50	1	73.00
Lead sheet	100	55.00	1	59.00
Stainless steel sheet	100	42.50	1	46.00
Zinc sheet	40	37.50	2	114.00
Clay tiles	40	33.00	10	113.00
Concrete tiles	30	12.50	10	85.00
Fibre cement tiles	30	24.50	12	134.00
Resin slates	30	28.00	12	149.00
Welsh slates	100	46.00	12	56.00
Stone slates	100	110.00	12	133.00

administration, architecture and the behavioural and engineering sciences' (Spedding & Holmes, 1994, p. 1).

In practice, facilities management seeks to ensure optimal use of buildings and their services, and the most appropriate conditions for its occupants, through a combination of decisions and actions. These may include:

- development of maintenance information systems
- maintenance of the building structure and fabric
- maintenance of the building services
- management of the site and ancillary buildings/structures
- monitoring of the conditions within the building (e.g. building management systems)
- provision and control of user facilities
- monitoring and control of user requirements
- staffing (e.g. security, caretaking, cleaning, groundstaff, resident engineer)
- business activities (e.g. accounting, purchasing, marketing, communications)

Initiatives such as occupancy cost appraisal and profiling (OCAP), as developed by the Building Performance Group, offered an integrated approach to predicting and optimising the occupancy costs of a building, being based on a durability audit, costed maintenance profile, energy audit and condition survey.

More recent techniques, such as whole-life performance and costing, can assist in producing more appropriate building solutions at a lower capital cost with reduced running costs. End of service life may be considered to

occur 'at the period of time after initiation when a building or its parts no longer meet the performance requirements and when physical failure is possible and/or when it is no longer practical or economic to continue with corrective maintenance' (BRE, 2006).

Limitations of existing buildings

Buildings have typically been designed and built to fulfil certain primary functions. When such buildings are no longer required for these activities and are being considered for new uses, there are various issues that may need to be considered.

Buildings erected before the mid-twentieth century would have been constructed using traditional materials and methods of construction. Such methods, and the facilities that these buildings offer, will generally be below the standards required of today's buildings, and attention will need to be given to upgrading services and facilities. Such work might include:

- provision of thermal insulation
- provision of acoustic insulation
- provision of damp-proofing
- improvements in levels of natural light and ventilation
- supply and/or re-routing of new and existing services
- provision of fire precautions
- works to satisfy statutory requirements (e.g. disabled access, means of escape)

In addition to these general issues, certain building types may pose particular limitations as a result of their location, construction, servicing or former usage (Fig. 7.2).

Finding the right use for a building

Successful utilisation of a building relies on a variety of factors, and is as much to do with finding the right user as it is with altering the structure, fabric and services of the building. It is therefore necessary to understand both the building and the potential market (including supply, demand and investment potential) when assessing the feasibility of a particular course of action.

In assessing a building for adaptation or re-use, various issues and options should be considered:

- location (e.g. vandalism, illegal entry, arson, marketability)

- Agricultural buildings
  - rural location (e.g. isolation, cost of new service installations)
  - partial redundancy (e.g. part of building complex remains unused)
  - single volume spaces (e.g. barns)
  - contamination (e.g. animal urine, fertilisers)
  - presence of plant and machinery
  - small or inadequate number of windows
  - specific details (e.g. opposing full-height doorways)
- Industrial buildings
  - depth of plan (e.g. reduced light and ventilation)
  - large voids (e.g. full height for machinery)
  - contamination (e.g. ground, floor surfaces)
  - presence of plant and machinery
  - restrictions on space (e.g. regular column positions)
  - limited headroom (e.g. tension rods to jack arch floors)
- Large domestic properties
  - external detailing (e.g. balconies, parapets)
  - internal fixtures and fittings
  - ancillary buildings (e.g. stables)
  - gardens and designed landscapes
  - garden features (e.g. fountains, boundary walls)
- Institutional
  - restrictive covenants
  - large spaces (e.g. assembly halls)
  - specific features of original use (e.g. high window sills in schools)
  - limitations of servicing (e.g. heating large spaces)
  - ancillary buildings
  - large gardens or grounds (e.g. playgrounds, parade grounds)
- Churches and chapels
  - lack of utility services
  - service connections (e.g. excavations in burial grounds)
  - restrictive covenants (e.g. sale and consumption of alcohol in church properties)
  - public rights of way (e.g. public footpaths)
  - public access (e.g. active burial grounds)
  - public feeling and resentment of change
  - external details (e.g. pinnacles, parapets, spires)
  - internal fixtures and fittings (e.g. bells, galleries, monumental brasses, pews, pulpits, wall paintings)
  - historic contents (e.g. tables, chairs, chests, books)

Fig. 7.2 Limitations associated with building types.

- sensitivity of use and/or occupancy (e.g. threats to staff, terrorism)
- statutory restrictions and requirements (e.g. planning policies, health and safety)
- legal considerations (e.g. easements, rights to light, party walls, rights of way)

- purchase/development opportunities (e.g. back-to-back deals, planning gain)
- form(s) of construction (e.g. inherent defects, material limitations)
- spatial configuration (i.e. geometry, open-plan, cellular)
- nature and extent of accommodation (e.g. volume, gross and net floor areas, floor-to-ceiling heights)
- future and ultimate potential for functional and/or organisational change (e.g. usable roof space, usable voids or spaces between buildings, vehicle parking, means of escape, structural limitations)
- potential for selective demolition to achieve overall viability
- hazards and associated risks to the health and safety of contractors and users (e.g. defective flooring, loose asbestos fibres)
- particular features worthy of protection and retention
- patterns of circulation (e.g. people, disabled persons, information, materials, goods)
- provision of services and associated equipment (e.g. raised floors, suspended ceilings, fixed work stations, 'hot desks', 'hotelling', cordless equipment)
- increased energy efficiency (e.g. environmental impact assessment, computerised building management, heat recovery, passive cooling and ventilation, photovoltaic glazing)
- improved working environment (e.g. air quality, natural lighting and ventilation)
- opportunities for environmentally friendly and sustainable options (e.g. waste management, rainwater harvesting, recycling of grey water)

Consideration should also be given to the particular needs and wishes of the client or potential building user, including:

- aspirations and/or expectations (i.e. what is expected from the building now and in the foreseeable future?)
- future lifestyle trends (e.g. increased car parking provision, home working, mixed tenure)
- functional requirements (e.g. space planning, floor loadings, servicing, partial occupation, property security)
- user requirements (e.g. ergonomics, disabled access and facilities, personnel security, comfort standards)
- finances (e.g. grants, loans, taxation, rents, outgoings)
- timescale (e.g. phased occupation, pre-lets)
- client (e.g. many individuals and organisations are taking an increasing interest in feng shui, with potential implications for internal layouts and decorative schemes)

Using historic buildings and sites

Buildings, monuments and areas of architectural or historic interest play an important part in the character and setting of urban and rural communities, and provide material continuity between each successive generation or phase of human development. There is thus an intimate relationship between these tangible reminders of the past and the needs and aspirations of today that has implications for education, employment, tourism, training, leisure and recreation.

There are also intangible elements to everyday life – including practices, representations, expressions, knowledge and skills, as well as associated instruments, objects, artefacts and cultural spaces – that are recognised by communities and groups. This intangible cultural heritage is often transmitted from generation to generation, and re-created over time to respond to changing circumstances. Safeguarding of this intangible heritage may include 'oral traditions and expressions, including language as a vehicle of the intangible cultural heritage; performing arts; social practices, rituals and festive events; knowledge and practices concerning nature and the universe; and traditional craftsmanship' (UNESCO, 2003).

Public attitudes toward the historic environment, and the current acceptance of conservation ideologies and constraints imposed on individuals and property owners, demonstrate a recognition of this powerful stimulus. Interest in, and care for, our historic past thus continues to grow and attract those who wish to contribute or simply to know more (Fig. 7.3).

The ways in which such historic buildings, monuments and areas are managed, and the tasks of repair and maintenance that are needed to keep them in an appropriate condition, have developed within a specific framework that encompasses many separate and related issues:

- philosophical (e.g. international charters, national amenity societies, local needs)
- technical (e.g. advisory services, professional bodies)
- legal (e.g. listed buildings, scheduled ancient monuments, conservation areas)
- financial (e.g. grant aid, taxation, investment potential)
- managerial (e.g. Heritage Partnership Agreements, Townscape Heritage Initiatives) (see Chapter 6)
- curatorial (e.g. quinquennial inspections, planned preventive maintenance)

These issues, and others, are enshrined in various international charters, resolutions, declarations and recommendations drawn up by the United Nations' Educational and Scientific Organisation (UNESCO), the Council

Fig. 7.3 The National Trust was founded in 1895 to 'promote the permanent preservation for the nation of land with outstanding natural features and animal and plant life, and buildings of beauty or historic interest'. Today it owns, manages and protects 252 497 hectares of land, 1135 km of coastline, 166 historic houses, 19 castles, 47 industrial monuments, 49 churches or chapels, 35 pubs and inns, over 230 gardens or landscape parks, and has 3.4 million members. In 2005–6 it attracted in excess of 12 million people to those of its properties that are open at a charge to the public (National Trust, 2006), including the modest semi-detached Edwardian property with its 1930s' interior known as Mr Straw's House in Worksop, Nottinghamshire (above).

of Europe and the International Council on Monuments and Sites (ICO-MOS), and in guidance documents and legal enactments produced by the individual countries. Some of the more influential and widely known within the United Kingdom include:

- *Manifesto* of Society for the Protection of Ancient Buildings (1877)
- Athens Charter (1931) – restoration of historic monuments
- Venice Charter (1966) – conservation and restoration of monuments and sites

- Florence Charter (1982) – preservation of historic gardens
- Planning (Listed Buildings and Conservation Areas) Act 1990
- DoE Planning Policy Guidance 16: Archaeology and Planning (1990)
- DoE/DNH Planning Policy Guidance 15: Planning and the Historic Environment (1994)
- British Standard 7913 (1998) – guide to the principles of the conservation of historic buildings
- Burra Charter (1999) – Australian charter for conservation of places of cultural significance
- English Heritage – *Power of Place: The Future of the Historic Environment* (2000)
- Department for Culture, Media and Sport – *The Historic Environment: A Force for Our Future* (2001)

Principles of building repair

The definition of 'repair' used at the beginning of this chapter – 'restoration of an item to an acceptable condition by the renewal, replacement or mending of worn, damaged or decayed parts' – highlights the breadth of work that may be undertaken to deal with defects, damage and decay as they affect building elements, components or individual materials (Fig. 7.4).

The principles of repair, which by necessity are closely allied to those of building maintenance (see below), are typically based on some or all of the following:

- complying with specific requirements (e.g. statutory, health and safety, lease or covenant obligations)
- satisfying functional, performance, statutory and/or user requirements
- removing or treating defects
- slowing rates of deterioration and decay
- safeguarding value and utility of building and facilities
- achieving desired or expected standards

In undertaking works of repair, it is advisable to:

- understand the building, and the nature and extent of proposed works, prior to commencement
- plan and execute works within a defined programme of inspection, repair and preventive maintenance (see below)

Fig. 7.4 Building repair requires an awareness and appreciation of the history, construction and materials of the building in order to achieve a satisfactory outcome.

- prioritise works to make best use of resources (e.g. urgent, necessary and desirable)
- execute works in a logical order to maximise use of resources (e.g. scaffolding) and minimise disruption to occupants (e.g. phased with occupation)
- satisfy health and safety needs of operatives and third parties (e.g. personal protective equipment, security)
- incorporate works to enhance the performance of the building (e.g. thermal insulation)
- incorporate works that will aid future maintenance and use of the building (e.g. improved roof access, *in-situ* monitoring systems)
- record nature and extent of actual works upon completion (e.g. health and safety file)

Defining what is an 'acceptable condition' will require consideration of the use and importance of the building, the demands of relevant legislation, and the needs and wishes of the owner or user. In the case of historic buildings and monuments, this needs to be set within the ethical context of current conservation theory and practice.

Repairing historic buildings and monuments

The management and aftercare of historic buildings, monuments and areas have to respect the framework of principles established by the documents and enactments noted above. Those that are relevant to the repair of buildings and monuments are summarised in Fig. 7.5.

Principles of building maintenance

Buildings and their services inevitably become obsolete as a result of factors relating to the use of the building (functional, economic, locational, social, statutory or physical) and changes in the needs and aspirations of the building user. The ageing and obsolescence of buildings, together with a corresponding depreciation in value, is therefore a continuous process (Fig. 7.6), but can be slowed or reversed by appropriate repair and maintenance.

Nature of maintenance

Maintenance may be undertaken either in anticipation of failure (preventive maintenance) or carried out to restore the building to an acceptable standard after failure (corrective maintenance). What is considered to be an 'acceptable standard' will be determined in relation to the importance of the building (is it, for instance, listed as being of special architectural or historic interest?), the building user and the use to which the building is put (Fig. 7.7).

The establishment of a programme of planned maintenance, which is typically both preventive and corrective in nature, has to be 'organised and carried out with forethought, control and the use of records to a predetermined plan' based on the results of previous condition surveys (BS 8210:1986). Such an approach has many benefits (Fig. 7.8), including:

- retaining the value of the property
- achieving optimum usage by minimising indirect maintenance costs
- presenting a good appearance

Purpose of repair
Primary purpose is to restrain the process of decay without damaging the character of the building, altering features that give historic or architectural importance, or unnecessarily disturbing or destroying historic fabric.

Need for repair
Intervention through repair must be kept to the minimum required to stabilise and conserve, with the aim of achieving a sufficiently sound structure to ensure long-term survival and meet the requirements of an appropriate use.

Avoiding unnecessary damage
The authenticity of a historic building depends crucially on its design and the integrity of its fabric. The unnecessary replacement of historic fabric ... will have an adverse effect on its appearance, will seriously diminish authenticity, and will significantly reduce its value as a source of historical information.

Analysing historic fabric
A thorough understanding of the historic development of a building is a necessary preliminary to its repair. Archaeological and architectural investigation, recording and interpretation and an assessment in the wider context may be required.

Analysing the causes of defects
Detailed design of the repairs should be proceeded by a survey of structural defects, and an investigation of the nature and condition of materials and the causes and processes of decay.

Adopting proven techniques
The aim should be to match existing materials and methods of construction in order to preserve appearance and historic integrity. New methods and techniques should only be used where they have proved themselves over a long period.

Truth to materials
Repairs should be executed honestly, with no attempt at disguise or artificial ageing, and not unnecessarily obtrusive or unsympathetic.

Removal of damaging alterations
Additions or alterations, including earlier repairs, are of importance in the cumulative history of the building ... strong presumption in favour of their retention.

Restoration of lost features
Some elements of a building that are important to its design ... may have been lost. Where of structural significance they will normally be replaced, but a programme of repair may also offer the opportunity for reinstatement of non-structural elements.

Safeguarding the future
A historic building should be regularly monitored and maintained, and where possible provided with an appropriate and sympathetic use.

Fig. 7.5 Principles of conservation repair (English Heritage, 1993; Brereton, 1995).

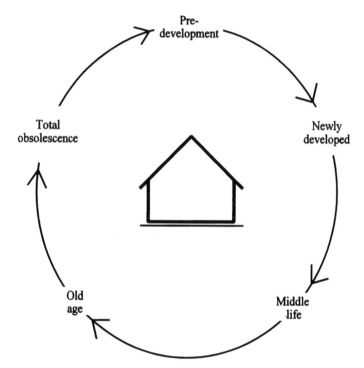

Fig. 7.6 Cycle of building obsolescence.

- maximising the life of the materials and components
- ensuring best use of materials and components
- maintaining user morale
- ensuring suitable standards of health, safety and security
- decreasing insurance risks
- ensuring compliance with regulations and Acts of Parliament

If overall maintenance is planned, preventive maintenance will reduce the costs of attending to emergencies and defects. In Flanders and the Netherlands, for instance, the system of 'Monumentenwacht' ensures that the historic buildings of its members are well maintained by using a travelling team of independent advisers and craftsmen to report on the condition of the buildings and carry out minor repairs (Binst, 1995; Dann & Worthing, 1998). Such an initiative has been piloted by Maintain our Heritage (2003) in Bath, and shown to have potential, subject to cost and market forces.

Planned preventive maintenance

The aim of a maintenance plan or programme is to maintain the building and site in an appropriate condition for a suitable period of time. The

Fig. 7.7   A lack of maintenance will result in accelerated deterioration and decay, with long-term consequences for the use of the building.

condition itself, being that regarded as appropriate for the building and its user, and the period, which may be related to the design life or other predetermined term, need to be carefully defined in order to balance satisfactorily the forms and costs of the maintenance undertaken.

A programme of preventive maintenance seeks to ensure that work is carried out at 'predetermined intervals, or corresponding to prescribed criteria, and intended to reduce the probability of failure or performance degradation of an item' (BS 3811:1993). As such, it must therefore be planned and constantly revised to take account of new information.

A long-term maintenance plan, such as undertaken over a four- or five-year period, will:

- determine the general level of expenditure to achieve the defined standards
- avoid large fluctuations in annual expenditure
- determine the optimum time for carrying out major repairs and improvements
- determine the structure and staffing of the maintenance organisation

Fig. 7.8   Careful cleaning of stone surfaces will assist in maintaining the fabric and improving the appearance of the building.

A medium-term or annual plan will provide a more accurate assessment of the amount of work to be carried out in the forthcoming year, and form the basis for the setting of financial budgets. The programme will build up from:

- work brought forward from the long-term plan as deemed necessary
- work disclosed by annual inspection
- work requested by users at the time of inspection
- an allowance for work requested by the users, but not capable of precise definition at the time of inspection
- an allowance for routine day-to-day maintenance based on past records

A short-term (for example, monthly) plan will develop from work brought forward from the medium-term plan, and should be sufficiently

detailed to give a sequence and duration for the works and a breakdown of requirements for labour, plant and materials.

Principles of preventive conservation

Preventive conservation, as distinct from preventive maintenance, may be defined as monitoring and controlling the main agents of destruction (including light, relative humidity, temperature, atmospheric pollutants, pests, accidents and disasters) to ensure the practical and cost-effective use and aftercare of sensitive or valuable buildings and their contents.

Although the principles of preventive conservation have typically been developed by those dealing with museums and galleries, they can equally well be applied to buildings and monuments, together with their fixtures and fittings. This is particularly so for those buildings and monuments that are empty or intermittently used (such as churches and chapels).

Practical considerations

In assessing a building or monument with a view to implementing a pro-gramme of preventive conservation, it is necessary to consider some, or all, of the following:

- location and situation (e.g. prevailing wind, sources of atmospheric pollution)
- building construction (e.g. levels of insulation, air infiltration, weath-ertightness)
- building morphology (e.g. volume, area, floor-to-ceiling heights)
- occupancy (e.g. constant, intermittent, empty)
- user activities (e.g. public access, floor loading)
- support activities (e.g. security, cleaning staff)
- environmental conditions (e.g. temperature, humidity, light, ventila-tion, fluctuations in conditions)
- risk assessments (e.g. theft of valuable objects)
- presence of contaminants and pollutants (e.g. chemical interaction and off-gassing from exhibits)
- documentation (e.g. inventories for fixtures, fittings and chattels)

Putting theory into practice

Once this information has been collected and collated, it is necessary to consider ways in which the building and contents may be protected from unnecessary risk or damage. In this, consideration should ideally be given

to ways in which existing facilities and services may be modified to improve conditions, before deciding on the addition or insertion of new installations.

Preventive conservation measures may include some or all of the following:

- modification of building fabric (e.g. draught sealing, additional thermal insulation)
- provision of physical buffers and/or barriers (e.g. storm porches, display cases, reordering use of rooms, changing routes around building)
- modification of user expectations (e.g. interpretation of reduced light levels)
- reduction of light levels (e.g. curtains, blinds, shutters, ultraviolet film/filters)
- revision to operation of existing building services (e.g. provision of constant background heat)
- modification of existing building services (e.g. use of humidistats instead of thermostats)
- installation of new building services (e.g. air conditioning, humidification, dehumidification)
- zoning of building (e.g. light-sensitive objects located in centre of building)
- procedures to deal with pollution (e.g. regular sampling, use of indicators)
- procedures to control insect infestations (e.g. trapping)

Planning for disasters and emergencies

Disasters, whether natural (fire, flood) or human (terrorism or vandalism), will inevitably affect both buildings and their owners, and have serious implications for those who work in, or are responsible for, the management and aftercare of such buildings (Fig. 7.9)

The concept of forward planning to take account of accidents and emergencies is increasingly been adopted by businesses, corporations and building users who are involved with, or responsible, for valuable objects or sensitive information (including museums, galleries and financial institutions).

Planning for an accident or disaster, and establishing a mechanism that will ensure swift and effective action, is a specialised skill that requires detailed knowledge of the building and its various occupants and users. It is dependent on integrated communication and organisation, and reliant on the coordination and support of owners, users and staff.

Fig. 7.9   Fire at Stoke Rochford Hall, Lincolnshire on 25 January 2005. (Photograph by Bob Stewart.)

Preparing a disaster or emergency plan

The preparation of a disaster or emergency plan is therefore based on a thorough inspection of the building and an assessment of the hazards and risks associated with its usage, together with a programme of training and familiarisation. Planning for a disaster will therefore include:

- regular inspection and survey of buildings and structures
- regular inspection and testing of building services
- preparation and maintenance of records (e.g. floor plans showing fire extinguishers)
- preparation of risk assessments (e.g. identify ways to remove hazard or prevent/minimise risk)
- assessment of housekeeping (e.g. regular cleaning regime, appropriate methods and products)
- assessment of support staff and services (e.g. security, caretakers, ground staff)
- monitoring of work conditions (e.g. staff facilities, meetings)
- management of building works (e.g. induction for new staff and contractors, hot-works permits)

- preparation of inventories (e.g. written and photographic records, security coding)
- sourcing of information (e.g. emergency contact numbers, access keys/codes, insurance documents, maintenance and operating manuals, local authority disaster plans)
- sourcing of help and assistance (e.g. fire service, police, local contacts, response staff)
- assessing availability of duplicate information and/or equipment (e.g. copies of important information, such as inventories, maintained off-site)
- assessing specific protection for valuable objects or sensitive information
- developing logistics of disaster plan (e.g. communication network, training and familiarisation for staff, assignment of key responsibilities, fire practices, evacuation procedures)
- preparation of instructions for seldom-performed tasks
- providing practical instruction (e.g. use of fire extinguishers, facilities for disabled persons)
- reviewing and monitoring procedures (e.g. learn from experience, update procedures, publish and disseminate relevant information)

Managing unoccupied buildings and sites

When a building becomes vacant, it is important to respond immediately to a number of issues with the aim of safeguarding its structure, fabric and services in the short term, whilst options are considered for longer-term usage. It is important to acknowledge that such properties are subject to a greater range of defects and damaging actions than occupied premises, and to consider some or all of the following issues:

- greater likelihood of theft (e.g. lead roof coverings, internal features, garden ornaments) (Fig. 7.10)
- greater potential risk of arson and vandalism
- greater incidence of illegal entry and squatting
- increased rate of deterioration and decay (e.g. fluctuations in temperature, reduced ventilation, conditions conducive to fungal and mould growth)
- possible inherent hazards (e.g. instability, defective fabric, previous uses)
- environmental concerns (e.g. pest control, communicable disease control, refuse collection)

Fig. 7.10 Damage caused to an eighteenth-century lead cistern in an attempt to drain the water and remove from the site.

- hazardous substances (e.g. asbestos, chemical residues, radiation sources, trade and other wastes)
- dangerous features (e.g. missing floorboards, unsafe stairways)
- contaminated land

A survey and assessment, together with a structured risk assessment, should be undertaken to identify potential problems and inform the ways in which the building and site are managed and secured. Detailed information regarding the protection of unoccupied buildings, including a management checklist devised to reduce the risk of attack, is provided by the Fire Protection Association (formerly Loss Prevention Council) (1996).

Action to manage and protect an unoccupied building might include some or all of the following (Fig. 7.11):

- making the building secure against unauthorised entry
- removing graffiti as it occurs
- leaving doors open to promote air movement
- clearing and sweeping flues to promote ventilation
- removing vegetation and other materials away from outside walls

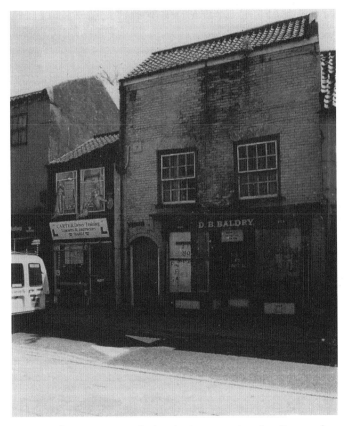

Fig. 7.11 Upper-floor accommodation in town centres is often underused and subject to deterioration and decay.

- removing all loose rubbish and readily combustible material (internal and external)
- regular sweeping, so that fresh rodent droppings are obvious at periodic inspections
- providing temporary support for vulnerable features (e.g. moulded plaster ceilings)
- removing valuable architectural features (e.g. fireplace surrounds, hardwood panelling) to a secure store (note potential requirement for listed building consent)
- terminating utility services (e.g. water, gas, electricity, telecommunications)
- draining down tanks, pipes, cisterns and boilers
- cancelling and redirecting postal service
- cancelling regular deliveries (e.g. milk, newspapers, groceries)
- advising emergency services (e.g. police and fire services)

Fig. 7.12   Keeping unoccupied buildings, of whatever age or construction, in good order requires careful planning and appropriate levels of protection.

- undertaking essential repairs
- regular inspecting/monitoring (e.g. vulnerable external fabric such as roof coverings, rainwater disposal system, doors, windows)
- initiating planned preventive maintenance (including gardens and grounds)
- providing security (e.g. alarms, boarding over openings, locks, lights, curtains, guards, caretakers)
- checking insurance cover (e.g. household cover lapses after specified period of non-occupancy)
- considering temporary uses (e.g. make use of attached living accommodation)

Short-, medium- or long-term 'moth-balling', which goes beyond simple and often potentially damaging measures such as boarding over windows and doors, should also be considered. Such an approach allows for recording and documentation, repairs and emergency works, and, in certain cases, the provision of background heating, mechanical ventilation and environmental monitoring (Mitchell, 1988; Park, 1993; Hutton & Lloyd, 1993). Whilst usually undertaken with historic buildings and monuments, the principles are equally applicable for any type of building (Fig. 7.12).

Fig. 7.13   How and where we live can have a significant effect on our health and well-being.

Health and the built environment

The World Health Organization's *European Charter on Environment and Health* states that every individual is entitled to 'an environment conducive to the highest attainable level of health and wellbeing' and 'the health of every individual, especially those in vulnerable and high-risk groups, must be protected' (WHO, 1989).

The question of health and the built environment is one that affects all of us, whether as users and occupiers of buildings or as those who work to repair and maintain them. Legislation can only affect and control direct action in and around buildings; it can do little to protect us from the indirect effects of poor design, specification and construction practices. Recent and continuing research has, however, raised a number of important issues and increased many people's awareness of the wider issues of health, sustainability and the environment (Fig. 7.13).

Unfit buildings

There are an estimated 6.3 million dwellings (out of a stock of 21.6 million, representing 29%) that fail to meet the government's decent homes standard in England (DCLG, 2006). This benchmark is based on the statutory minimum standard for housing, a reasonable state of repair, having reasonably modern facilities and services, and providing a reasonable degree of thermal comfort. Those most often failing this standard are purpose-built high-rise flats (51.5%) and houses built before 1919 (42.4%), with the most common reason for failure relating to thermal comfort (i.e. lack of effective insulation or efficient heating). The average cost of bringing a dwelling up to a decent standard is £6650, depending on the criteria on which it failed. Homes that fail on thermal comfort require £1884, whilst those in need of work to meet other criteria require £13 508.

A dwelling has previously been considered 'unfit' under the Housing Act 1985 if it was not structurally stable, free from serious disrepair and free from dampness that threatens the health of the occupants. Each dwelling must also have had:

- adequate lighting, heating and ventilation
- an effective drainage system
- a suitably located toilet for the exclusive use of the occupants
- a suitably located bath or shower and basin, each with a proper supply of hot and cold water
- satisfactory facilities for the preparation and cooking of food, including a sink with a proper supply of hot and cold water

Since April 2006 this housing fitness regimen has been replaced by a risk assessment procedure – the Housing, Health and Safety Rating System (HHSRS) – introduced by the Housing Act 2004. HHSRS assesses 29 categories of housing hazard categorised under physiological requirements (e.g. damp and mould growth, excess cold, biocides), psychological requirements (e.g. crowding and space, lighting, noise), protection against infection (e.g. domestic hygiene, pests, refuse, food safety, water supply), and protection against accidents (e.g. falling, electrical hazards, structural collapse and failing elements), provides a rating for each hazard so that the assessment is based on the risk to the potential occupant who is most vulnerable to that particular hazard.

Bad housing conditions have a demonstrable effect on the health, safety and well-being of occupants, particularly children and those who are elderly or in poor health. The effects include damp and mould, inadequate heating and ventilation, increasing incidence of asthma, childhood accidents resulting from badly designed housing and dangerous fittings, illness

resulting from indoor pollutants, and fuel poverty (Shelter, 1997; 2005; 2006). There are also significant issues relating to overcrowded housing, living in temporary accommodation, and limited housing in rural areas.

Health and safety legislation

A large and complex body of health and safety law (both common and statute) assigns rights and responsibilities to all parties involved in construction, including clients, designers, suppliers, landlords and contractors. Criminal and civil liabilities may arise as a result of a breach of health and safety law. The most important piece of legislation dealing with occupational health and safety is the Health and Safety at Work, etc Act 1974, which applies to every type of work situation. The Act sets out general duties for the health and safety of those involved in work, including employers, employees, the self-employed, suppliers of work equipment and those who control work premises.

Although not falling within the main body of health and safety legislation, the Occupiers' Liability Acts 1957 and 1984 are relevant to occupiers of premises and their responsibilities towards contractors working on the premises and trespassers. Since the European Communities Act 1972, the United Kingdom is required to comply with all European Union legislation. The implementation of health and safety legislation in Great Britain is covered by the Management of Health and Safety at Work Regulations 1999, as amended, whilst specific regulations are in force covering health, safety and welfare in the workplace (see below); the provision and use of work equipment; personal protective equipment at work; manual handling operations; and display screen equipment.

The Workplace (Health, Safety and Welfare) Regulations 1992 apply to all workplaces, other than those used for construction, mining and certain other activities, and require employers and others in control of workplaces to ensure appropriate levels of maintenance, ventilation, indoor temperatures, lighting, cleanliness, space provision, sanitary conveniences, washing facilities and drinking water.

Construction regulations covering health and safety have been consolidated into the Construction (Health, Safety and Welfare) Regulations 1996, which covers activities such as setting up the site, earth moving, demolition, excavations and foundations, temporary access, erection of superstructure and structural stability, access and permanent works, fire precautions, emergency procedures, training and supervision, and requirements for inspection and reporting.

The Reporting of Injuries, Diseases and Dangerous Occurrences Regulation 1995 (RIDDOR) require employers, self-employed and those

responsible for work premises to report deaths, major injuries (e.g. amputation, dislocation, loss of sight), over-three-day injuries, diseases (e.g. poisonings, skin diseases, lung diseases, infections) and dangerous occurrences (e.g. explosion, collapse or bursting of vessels and pipes) to the Incident Contact Centre (Caerphilly), local authority environmental health department, or HSE area office.

The Control of Substances Hazardous to Health Regulations (COSHH) 2002 apply to all substances capable of causing adverse health effects, and include chemicals, biological agents, carcinogens, dusts and allergens. The COSHH regulations require employers to assess risks posed by exposure to hazardous substances in the workplace; consider what precautions are needed; prevent, or at least adequately control, such exposure; provide, maintain, test and examine suitable control measures and ensure that they are used; monitor workplace exposure against prescribed exposure limits; provide health surveillance; prepare plans and procedures to deal with accidents, incidents and emergencies; and provide relevant information, instruction and training to employees.

The Construction (Design and Management) Regulations 1994, as amended, have brought about a significant change in the ways in which health and safety are considered in relation to construction work. There is now a duty on the client, designer and principal contractor to consider how work will be managed on site and to ensure that certain requirements are met before, during and after the work is undertaken.

The CDM regulations apply to any project involving notifiable construction work (i.e. the construction phase will be longer than 30 days or will involve more than 500 person-days of construction work) or demolitions, and to projects that are not notifiable, but where there will be five or more persons involved at any one time. Construction work is considered to include fitting out, commissioning, upkeep, redecoration, maintenance, site preparation and demolition. The revised Construction (Design and Management) Regulations, implemented in Spring 2007, are intended to simplify existing provisions and bring together the existing CDM regulations and the Construction (Health, Safety and Welfare) Regulations 1996 into a single regulatory package.

We do not live or work in a risk-free society, and so it is necessary for those involved in construction to consider the likely hazards that are to be faced when working on a site, using the building or maintaining its facilities. Risk assessment has therefore become an important part of the construction process and should be used to identify hazards, assess the severity of associated risks, and indicate what action should be taken to remove or reduce the likelihood of injury.

Alternative approaches to management and aftercare

Growing concern for the environment and the health of those living in, or working on, buildings has led to changes in the way we perceive and deal with buildings and building defects. As a result, the conventional approach to material selection and remediation has had to be rethought and placed in the context of a wider environmental agenda.

The retention and adaptation of existing buildings is often the most cost-effective and sustainable way in which to meet society's demands for housing and for places of leisure, recreation and work, yet potentially viable buildings are often demolished to make way for new prestigious buildings that offer high returns for financiers and shareholders. Such profligacy is, however, becoming unacceptable as the residual value of the national building stock is increasingly recognised and safeguarded.

Adaptation of existing buildings for new uses is becoming an economic and social necessity, which requires a reassessment of the role and value of traditional buildings and areas; this is particularly the case with those listed as being of special historic or historic interest. It is not possible to continue protecting and legislating against the changes that are needed to bring such buildings back into beneficial use. Sensible and sensitive schemes, which combine accepted conservation theory with the best of today's design practice, are needed to move us towards a viable future (Fig. 7.14) (see Chapter 6).

At the level of individual building projects, the specification and use of alternative materials to replace those considered to be environmentally damaging or hazardous to health, including those that appear in lists of deleterious materials in collateral warranties or as conditions in professional contracts (BRE Digest 425, 1997; Ruston, 2006), continues to become more widely practised. Materials with short life-spans and high embodied energies are increasingly penalised or prohibited by environmentally conscious clients, and eco-labelling of building materials and products is gaining acceptance (Fig. 7.15) (BRE IP 11, 1993). Information on products and services is also widely available, including publications such as *The Green Guide to Specification* (Anderson *et al.*, 2002); *The Whole House Book: Ecological Building Design and Materials* (Harris & Borer, 2005); *Strategies for Sustainable Architecture* (Sassi, 2006); and *The Green Building Bible* (Vols 1 and 2) (Hall, 2006).

Issues of health and comfort also increasingly require consideration, by those responsible for building management and aftercare. Poor working conditions have, for instance, been shown to have a significant effect on efficiency and productivity, whilst increases in asthma and respiratory diseases have in certain cases been linked to building defects and deficiencies.

Fig. 7.14   The reuse of former industrial buildings, such as at Saltaire in West Yorkshire, can provide flexible accommodation for a variety of uses.

Such problems require a combination of actions, based on careful planning and implementation, that require a shift in current thinking and practice (Singh, 1996, p. 30). Such an approach might include:

- removing the source of pollution (e.g. ban smoking)
- avoiding particular pollutants (e.g. use alternative materials such as low-solvent or solvent-free paints)
- isolating the source(s) of contamination or pollution (e.g. containment, encapsulation, shielding, sealing)
- designing to avoid problems (e.g. effective ventilation, thermal comfort, lighting, maintenance)
- removing existing contamination or reservoirs of pollutants (e.g. regular cleaning of furnishings)
- assessing and monitoring conditions (e.g. regular inspections or surveys)

Taking a more benign approach to managing decay in buildings is also possible, but requires a greater understanding of the mechanisms of deterioration and decay in order to be effective and reliable. 'Environmental' or 'green' approaches to the treatment of fungal and insect attack, for

Fig. 7.15   The EcoHouse in Leicester has exhibits and installations to show how it is possible to save energy and resources, and create a healthy environment, in and around the home.

instance, which make use of enhanced ventilation and improved preventive maintenance instead of the application of toxic chemicals, demonstrate that success is possible (Hutton *et al.*, 1991; Singh & White, 1995; Ridout, 1998).

Issues of sustainability and sustainable development

Sustainability has many interpretations, particularly when used in the contexts of both natural and built environments. Definitions such as 'improving the quality of human life while living within the carrying capacity of supporting ecosystems' (IUCN/UNEP/WWF, 1991, p. 10) and 'increasing the quality of life as well as preserving the environment, both for ourselves and future generations' (Woolley, 1997) demonstrate the broad application of the principles of sustainability, as established at the Earth Summit in Rio de Janeiro in 1992.

Current concerns, such as changing weather patterns (including the influences of global warming), pollution, reductions in poverty and social exclusion, demand for new housing (4.4 million new homes supposedly

needed by 2016), development of greenbelt land, use of brownfield sites and need for integrated public transportation), can all have a potential effect on our lives, and so need to be fully addressed and understood in advance of potentially harmful action.

Sustainable development (referred to as Agenda 21), which may be defined as 'development that meets the needs of the present without compromising the ability of future generations to meet their own needs' (Brundtland Commission, 1987) or '. . . achieving economic development in the form of higher living standards while protecting and enhancing the environment, and making sure that these economic and environmental benefits are available to everyone, not just the privileged few' (DETR, 1998), takes these principles further by considering the impact of development on the natural environment. This requires a structured approach to managing the needs of present and future generations with regard to housing, roads, energy, agriculture and forestry, and, in its widest sense, includes how we manage our existing stock of buildings.

In the UK, four priority areas for immediate action have been identified by government: sustainable consumption and production; climate change and energy; natural resource protection and environmental enhancement; and sustainable communities – with a common aim of changing the behaviour of individuals, communities, firms and the public sector set against international, regional and local indicators (DEFRA, 2006).

The consumption of finite resources for these activities also needs to be managed to make best use of what is available. In our lifetimes, for instance, each of us will typically account for 500 tonnes of sand, gravel, limestone and clay in the manufacture of concrete and bricks (Zalasiewicz, 1998, p. 28). Between 80% and 90% of all household and commercial waste from 56 million people in England and Wales currently goes into landfill sites (Gray, 1998, p. 173). The cost of dealing with 100 000–200 000 hectares of potentially contaminated land in United Kingdom is estimated to be £2 billion (Nathanail & Nathanail, 1998, p. 73). Is this the best use of our natural resources, and how should we plan for the future?

Current initiatives – including green taxes (e.g. aviation policy, fuel tax, climate-change levy), reductions in carbon emissions relative to Kyoto targets and energy consumption (e.g. transport systems, carbon sequestration), development of low-carbon technologies and alternative or renewable energy production (e.g. energy harvesting), recycling (e.g. Waste Electrical and Electronic Equipment directive, design for disassembly), and more rigorous standards for energy efficiency and performance (e.g. domestic energy rating) – will all help in achieving a sustainable future, but only if there is consensus, commitment and delivery.

Stewardship and assessments of value

At a general level, sustainability is concerned with issues of stewardship. This requires both knowledge of the asset or resource being managed, and an assessment of its present and future value. In the case of the historic environment (embracing archaeological sites, buildings, areas and the material remains of past human activities), English Heritage (2006) considers value under the following headings:

- *aesthetic* – relating to the ways in which people respond to a place through sensory and intellectual experience of it (e.g. townscapes, vernacular traditions, local building materials, designed and natural landscapes)
- *community* – relating to the meanings of a place for the people who identify with it, and whose collective memory it holds (e.g. evolution of society, understanding cultural roots)
- *evidential* – relating to the potential of a place to yield primary information about past human activity (e.g. development of past cultures)
- *historical* – relating to the ways in which a place can provide direct links to past people, events and aspects of life (e.g. context for life)
- *instrumental* – economic, educational, recreational and other benefits which exist as a consequence of the cultural or natural heritage values of a place (e.g. economic development, tourism, quality of life)

Such values are inherent in the development of conservation principles, policy and guidance for the sustainable management of the historic environment. The key principles on which this might be achieved are (English Heritage, 2006):

- the historic environment is a shared resource
- it is essential to understand and sustain what is valuable in the historic environment
- everyone can make a contribution
- understanding the values of places is essential
- places should be managed to sustain their significances
- decisions about change must be reasonable and transparent
- it is essential to document and learn from decisions

Value may also come from the symbiosis of built and natural forms. Integration of policies dealing with the natural and historical aspects of the countryside can, for example, show benefits when historical and organisational constraints are put aside, and information is shared between those responsible for managing and implementing change. It is this broadening

Fig. 7.16   The continued use of traditional building materials will ensure that supplies and associated craft skills remain in demand.

of outlook and interest in the wider issues that provides the challenges for those dealing with buildings and their settings.

Local character and distinctiveness, whether it be of a particular town, village or area of countryside, need also to be acknowledged and valued for their contribution to the overall diversity of an area or region (Fig. 7.16). Initiatives such as Local Agenda 21, which encourages individual and community participation in sustainable development and the protection of the environment at a grass-roots level, and the work of organisations such as Common Ground, assist in recognising and respecting these often fragile resources.

Buildings for the present and the future

In the final analysis, what do people expect and deserve from their buildings? Individuals want homes that are secure, comfortable and easy to look after. Tenants look for economy, improved productivity and adaptability to meet their changing needs. Investors seek improved financial returns. Society demands buildings that respect the environment and make a positive contribution to the community.

Buildings and their users need, however, to be better understood if they are to be successfully managed and cared for. Such an understanding will come by adopting the principles and practice of building pathology described in this book.

References

Anderson, J., Shiers, D. & Sinclair, M. (2002) *The Green Guide to Specification*, 3rd ed. Oxford: Blackwell Publishing.

Australia ICOMOS (1999) *The Burra Charter: The Australia ICOMOS Charter for the Conservation of Places of Cultural Significance.* Australia: Australia ICOMOS.

Baines, F. (1923) Preservation of ancient monuments and historic buildings. *RIBA Journal*, **XXXI** (4), 22 December, 104–6.

Binst, S. (1995) Monument watch in flanders and the Netherlands. In *The Economics of Architectural Conservation* (P. Burman, R. Pickard & S. Taylor eds), pp. 103–7. York: Institute of Advanced Architectural Studies.

Brereton, C. (1995) *The Repair of Historic Buildings: Advice on Principles and Methods*, 2nd ed. London: English Heritage.

British Standards Institution (1986) *British Standard Guide to Building Maintenance Management*. BS 8210. London: BSI.

British Standards Institution (1993) *Glossary of Terms used in Terotechnology*, 4th edn. BS 3811. London: BSI.

British Standards Institution (1998) *Guide to the Principles of the Conservation of Historic Buildings.* BS 7913. London: S1.

British Standards Institution (n.d.) *Guide to the Care of Historic Buildings.* Draft document. London: BSI.

Brundtland Commission (1987) *Our Common Future: The Report of the World Commission on Environment and Development.* Oxford: Oxford University Press.

Building Research Establishment (1993) *Ecolabelling of Building Materials and Building Products.* Information Paper 11. Garston: BRE.

Building Research Establishment (1997) *Lists of Excluded Materials: A Changing Practice.* Digest 425. Garston: BRE.

Building Research Establishment (2006) *Green Guide Update: BRE Response to Comments on Whole Life Performance.* Briefing Note (6). Garston: BRE.

Dann, N. & Worthing, D. (1998) How to ensure good conservation through good maintenance. *Chartered Surveyor Monthly*, **7** (6), 40–41.

Department for Communities and Local Government (2006) *English House Condition Survey 2004: Annual Report.* Wetherby: DCLG Publications.

Department for Environment, Farming and Rural Affairs (2006) *Sustainable Development: The Government's Approach – Providing Guidance on How to Pursue a More Sustainable Future.* London: DEFRA (Sustainable Development Unit).

Department of the Environment, Transport and the Regions (1998) *Sustainable Production and Use of Chemicals: Consultation Paper on Chemicals in the Environment.* London: DETR.

Douglas, J. (2006) *Building Adaptation*, 2nd ed. Oxford: Butterworth-Heinemann.

Duffy, F. (1990) Measuring building performance. *Facilities*, **8** (5), 17–20.

English Heritage (1993) *Principles of Repair.* London: English Heritage.

English Heritage (2000) *Power of Place: The Future of the Historic Environment.* London: English Heritage.

English Heritage (2006) *Conservation Principles for the Sustainable Management of the Historic Environment.* London: English Heritage.

Feilden, B.M. (2003) *Conservation of Historic Buildings*, 3nd edn. Oxford: Architectural Press.

Gray, J.M. (1998) Hills of waste: a policy conflict in environmental geology. In *Issues in Environmental Geology: A British Perspective* (M. Bennett, & P. Doyle eds), pp. 173–95. Bath: Geological Society Publishing House.

Hall, K. (ed). (2006) *The Green Building Bible* (Vols 1 and 2), 3rd edn. Llandysul: Green Building Press.

Harris, C. & Borer, P. (2005) *The Whole House Book: Ecological Building Design and Materials.* 2nd edn. Machynlleth: CAT Publications.

Hutton, T., Lloyd, H. & Singh, J. (1991) The environmental control of timber decay. *Structural Survey*, **10** (1), 5–20.

Hutton, T. & Lloyd, H. (1993) 'Mothballing' buildings: proactive maintenance and conservation on a reduced budget. *Structural Survey*, **11** (4), 335–42.

International Council on Monuments and Sites (1964) *Venice Charter for the Conservation and Restoration of Monuments and Sites*, Article 5, Venice: ICOMOS.

IUCN/UNEP/WWF (1991) *Caring for the Earth: A Strategy for Sustainable Living.* London: Earthscan.

Loss Prevention Council (1996) *Code of Practice for the Protection of Unoccupied Buildings*, 2nd edn. Borehamwood: Loss Prevention Council.

Maintain our Heritage (2003) *Historic Building Maintenance: A Pilot Inspection Service.* Bath: Maintain our Heritage.

Maintain our Heritage (2004) *Putting it Off: How Lack of Maintenance Fails Our Heritage.* Bath: Maintain our Heritage.

Macdonald, S. Hughes, T., Wood, C. & Strange, P. (2003) Saving England's stone, slate roofs: A model for the revival and enhancement of the stone slate roofing industry in the South Pennines. In *English Heritage Research Transactions – Volume 9: Stone Roofings* (C. Wood ed), pp. 1–31. London: James and James.

Mitchell, E. (1988) *Emergency Repairs for Historic Buildings.* London: Butterworth Architecture.

Nathanail, C.P. & Nathanail, J.F. (1998) Professional training for geologists in contaminated land management. In *Contaminated Land and Groundwater: Future Directions* (D.N. Lerner, & N.R.G. Walton eds), pp. 73–77. Engineering Geology Special Publication 14. London: Geological Society.

National Trust (2006) *2005/06 Annual Report and Financial Statements.* Swindon: National Trust.

Park, S. (1993) *Mothballing Historic Buildings.* Preservation Brief 31. Washington: US Department of the Interior, National Parks Service (Preservation Assistance).

Ridout, B. (1998) The treatment of timber decay into the 21st century. *Journal of Architectural Conservation*, **4** (3), 7–19.

Ruston, T. (2006) *Investigating Hazardous and Deleterious Materials.* Coventry: RICS Books.

Sassi, P. (2006) *Strategies for Sustainable Architecture.* Oxford: Taylor & Francis.

Shelter (1997) *Homelessness and Health: How Homelessness and Bad Housing Impact on Physical Health.* London: Shelter Information.

Shelter (2005) *Generation Squalor: Shelter's National Investigation into the Housing Crisis*. London: Shelter Information.

Shelter (2006) *Chance of a Lifetime: The Impact of Bad Housing on Children's Health*. London: Shelter.

Singh, J. (1996) Health, comfort and productivity in the indoor environment. *Indoor Built Environment*, **5**, 22–33.

Singh, J. & White, N. (1995) Dry rot and building decay: A greener approach. *Construction Repair*, **9** (2), 28–32.

Spedding, A. & Holmes, R. (1994) Facilities management. In *CIOB Handbook of Facilities Management* (A. Spedding ed), pp. 1–8. Harlow: Longman Scientific & Technical.

UNESCO (2003) *Convention for the Safeguarding of the Intangible Cultural Heritage*. Paris: UNESCO.

Woolley, N. (1997) RICS research evaluates the true cost of sustainability. *Chartered Surveyor Monthly*, **7** (3), November/December, 35.

World Health Organization (1989) *European Charter on Environment and Health*. Copenhagen: WHO.

Zalasiewicz, J. (1998) Buried treasure. *New Scientist*, **159** (2140), 27–30.

Further reading

Alexander, D. (2002) *Principles of Emergency Planning and Management*. Harpenden: Terra Publishing.

Anink, D., Boonstra, C. & Mak, J. (1996) *Handbook of Sustainable Building: An Environmental Preference Method for Selection of Materials for Use in Construction and Refurbishment*. London: James & James (Science Publishers).

Asif, M., Muneer, T. & Kelley, R. (2006) Life cycle assessment: a case study of a dwelling home in Scotland. *Building and Environment*, **42** (3), 1391–94.

Baker, D. & Meeson, B. (1997) *Analysis and Recording for the Conservation and Control of Works to Historic Buildings*. Chelmsford: Association of Local Government Archaeological Officers.

Bell, D. (1997) *The Historic Scotland Guide to International Conservation Charters*. Edinburgh: Historic Scotland.

Bevan, S. & Dobie, A. (2006) Save our Heritage. *RICS Business*, February, 16–20.

Blacker, J. (1997) *The Building Centre Maintenance Manual + Health and Safety File*. 5th ed. London: The Building Centre.

Building Employers Confederation/Federation of Master Builders (1994) *English Homes: A National Asset?* London: BCE/FMB.

Burman, P. (ed) (1994) *Treasures on Earth: A Good Housekeeping Guide to Churches and their Contents*. London: Donhead Publishing.

Chanter, B. & Swallow, P.G. (1996) *Building Maintenance Management*. Oxford: Blackwell Science.

Collings, J. (2002) *Old House Care and Repair*. London: Donhead Publishing.

Cunnington, P. (1988) *Change of Use: The Conversion of Old Buildings*. London: A. & C. Black (Publishers).

Curwell, S.R. & March, C.G. (eds) (1986) *Hazardous Building Materials: A Guide to the Selection of Alternatives*. London: E. & F.N.

Department for Culture, Media and Sport (2001) *The Historic Environment: A Force for Our Future*. London: DCMS (Architecture and Historic Environment Division).

Earl, J. (1996) *Philosophy of Building Conservation*. Reading: College of Estate Management.

Edwards, B. (1998) *Green Buildings Pay*. London: Routledge.

English Heritage (1996) *Sustainability and the Historic Environment*. Report prepared by Land Use Consultants and CAG Consultants. London: English Heritage.

English Heritage (1997) *Sustaining the Historic Environment: New Perspectives on the Future*. London: English Heritage.

Griffin, C. (ed) (1994) *No Losers – New Uses: New Homers from Empty Properties*. London: National Housing and Town Planning Council/Empty Homes Agency.

Hall, K. (ed) (2005) *The Green Building Bible*, 2nd edn. Llandysul: Green Building Press.

Hall, K. & Warm, P. (1998) *Greener Building: Products and Services Directory*, 4th edn. East Meon: AECB.

Harland, E. (1998) *Eco-Renovation: The Ecological Home Improvement Guide*, 2nd edn. Dartington: Green Books.

Highfield, D. (1987) *Rehabilitation and Re-Use of Old Buildings*. London: E. & F.N. Spon.

Highfield, D. (1991) *The Construction of New Buildings Behind Historic Facades*. London: E. & F.N. Spon.

Holdsworth, B. & Sealey, A. (1992) *Healthy Buildings: A Design Primer for a Living Environment*. Harlow: Longman.

Kemp, D. (1998) *The Environment Dictionary*. London: Routledge.

Laing, A., Duffy, F., Jaunzens, D. & Willis, S. (1998) *New Environments for Working: The Re-design of Offices and Environmental Systems for New Ways of Working*. London: E. & F.N. Spon.

Latham, D. (1998) *Creative Re-Use of Buildings*. Shaftesbury: Donhead Publishing.

Local Government Management Board (1997) *Local Agenda 21 in the UK – The First 5 Years*. London: LGMB.

Macdonald, S. (ed) (1996) *Modern Matters: Principles and Practice in Conserving Recent Architecture*. Shaftesbury: Donhead Publishing.

Matulionis, R.C. & Freitag, J.C. (eds) (1991) *Preventive Maintenance of Buildings*. London: Chapman & Hall.

Mills, E. (1994) *Building Maintenance and Preservation: A Guide to Design and Management*. 2nd edn. Oxford: Butterworth Architecture.

National Trust (2005) *The National Trust Manual of Housekeeping: The Care of Collections in Historic Houses Open to the Public*. Oxford: Elsevier.

Park, A. (1998) *Facilities Management: An Explanation*, 2nd edn. Basingstoke: Macmillan Press.

Pickard, R.D. (1996) *Conservation in the Built Environment*. Harlow: Addison Wesley Longman.

Pout, C., Moss, S. & Davidson, P. (1998) *Non-Domestic Building Energy Fact File*. Garston: BRE/DETR.

Roaf, S., Fuentes, M. & Thomas, S. (2005) *Ecohouse 2: A Design Guide*, 2nd edn. Oxford: Architectural Press.

Royal Institution of Chartered Surveyors (1996) *The Principles of Building Conservation.* Building Conservation Note No. 6. London: RICS.

Royal Institute of Chartered Surveyors (2007) *Disaster Management Process Protocol.* London: RICS/University of Salford.

Sadler, R. & Ward, K. (1992) *Owner Occupiers Attitudes to House Repairs and Maintenance.* London: Building Conservation Trust (Upkeep).

Seeley, I. (1987) *Building Maintenance,* 2nd edn. London: Macmillan Press.

Shiers, D., Howard, N. & Sinclair, M. (1998) *The Green Guide to Specification,* 2nd edn. CRC/BRE: Garston.

Son, L.H. & Yuen, G.C.S. (1993) *Building Maintenance Technology.* Basingstoke: Macmillan Press.

Spedding, A. (ed) (1994) *CIOB Handbook of Facilities Management.* Harlow: Longman Scientific & Technical.

Swallow, P. (1997) Managing unoccupied buildings and sites. *Structural Survey,* **15** (2), 74–79.

Thomas, R. (ed.) (1996) *Environmental Design: An Introduction for Architects and Engineers.* London: Chapman & Hall.

Vale, B. & Vale, R. (1991) *Green Architecture: Design for a Sustainable Future.* London: Thames & Hudson.

van Wagenberg, A.F. (1997) Facility management as a profession and academic field. *International Journal of Facilities Management,* **1** (1), 3–10.

Watt, D. & Colston, B. (eds) (2003) *Conservation of Historic Buildings and their Contents.* Shaftesbury: Donhead Publishing.

Wood, B. (2003) *Building Care.* Oxford: Blackwell Science.

Woolley, T. & Kimmins, S. (2002) *Green Building Handbook: Volume 2.* London: Spon Press.

Woolley, T., Kimmins, S., Harrison, P. & Harrison, R. (2005) *Green Building Handbook.* Abingdon: Taylor & Francis.

Wordsworth, P. (2003) *Lee's Building Maintenance Management.* Oxford: Blackwell Publishing.

Appendix A

Requirements of Schedule 1 to the Building Regulations 2000

A STRUCTURE (2004 EDITION)

A1 LOADING

(1) The building shall be constructed so that the combined dead, imposed and wind loads are sustained and transmitted by it to the ground (a) safely, and (b) without causing such deflection or deformation of any part of the building, or such movement of the ground, as will impair the stability of any part of another building.

(2) In assessing whether a building complies with (1) regard shall be had to the imposed and wind loads to which it is likely to be subjected in the ordinary course of its use for the purpose for which it is intended.

A2 GROUND MOVEMENT

The building shall be constructed so that ground movement caused by (a) swelling, shrinkage or freezing of the subsoil, or (b) land-slip or subsidence (other than subsidence arising from shrinkage), in so far as the risk can be reasonably foreseen, will not impair the stability of any part of the building.

A3 DISPROPORTIONATE COLLAPSE

The building shall be constructed so that in the event of an accident the building will not collapse to an extent disproportionate to the cause.

B FIRE SAFETY (2000 EDITION, INCORPORATING 2000 AND 2002 AMENDMENTS)

B1 MEANS OF WARNING AND ESCAPE

The building shall be designed and constructed so that there are adequate provisions for the early warning of fire, and appropriate means of escape

in case of fire from the building to a place of safety outside the building capable of being safely and effectively used at all material times.

B2   INTERNAL FIRE SPREAD (LININGS)

(1) To inhibit the spread of fire within the building, the internal lining shall (a) adequately resist the spread of flame over their surfaces, and (b) have, if ignited, either a rate of heat release or a rate of fire growth, which is reasonable in the circumstances.

(2) In this paragraph 'internal linings' mean the materials or products used in lining any partition, wall, ceiling or other internal structure.

B3   INTERNAL FIRE SPREAD (STRUCTURE)

(1) The building shall be designed and constructed so that, in the event of fire, its stability will be maintained for a reasonable period.

(2) A wall common to two or more buildings shall be designed and constructed so that it adequately resists the spread of fire between those buildings. For the purposes of this sub-paragraph a house in a terrace and a semi-detached house are each to be treated as a separate building.

(3) To inhibit the spread of fire within the building, it shall be sub-divided with fire-resisting construction to an extent appropriate to the size and intended use of the building.

(4) The building shall be designed and constructed so that the unseen spread of fire and smoke within concealed spaces in its structure and fabric is inhibited.

B4   EXTERNAL FIRE SPREAD

(1) The external walls of the building shall adequately resist the spread of fire over the walls and from one building to another, having regard to the height, use and position of the building.

(2) The roof of the building shall resist the spread of fire over the roof and from one building to another, having regard to the use and position of the building.

B5   ACCESS AND FACILITIES FOR THE FIRE SERVICE

(1) The building shall be designed and constructed so as to provide reasonable facilities to assist fire fighters in the protection of life.

(2) Reasonable provision shall be made within the site of the building to enable fire appliances to gain access to the building.

C   SITE PREPARATION AND RESISTANCE TO CONTAMINANTS AND MOISTURE (2004 EDITION)

C1   PREPARATION OF SITE AND RESISTANCE TO CONTAMINANTS

    (1)   The ground to be covered by the building shall be reasonably free from any material that might damage the building or affect its stability, including vegetable matter, topsoil and pre-existing foundations.

    (2)   Reasonable precautions shall be taken to avoid danger to health and safety caused by contaminants on or in the ground covered, or to be covered by the building and any land associated with the building.

    (3)   Adequate sub-soil drainage shall be provided if it is needed to avoid:
    (a) the passage of ground moisture to the interior of the building; or
    (b) damage to the building, including damage through the transport of water-borne contaminants to the foundations of the building.

    (4)   For the purposes of this requirement, "contaminant" means any substance which is or may become harmful to persons or buildings including substances, which are corrosive, explosive, flammable, radioactive or toxic.

C2   RESISTANCE TO MOISTURE

The floors, walls and roof of the building shall adequately protect the building and people who use the building from harmful effects caused by (a) ground moisture; (b) precipitation and wind-driven spray; (c) interstitial and surface condensation; and (d) spillage of water from or associated with sanitary fittings or fixed appliances.

D   TOXIC SUBSTANCES (1992 EDITION, AMENDED 2002)

D1   CAVITY INSULATION

If insulating material is inserted into a cavity in a cavity wall reasonable precautions shall be taken to prevent the subsequent permeation of any toxic fumes from that material into any part of the building occupied by people.

E   RESISTANCE TO THE PASSAGE OF SOUND (2003 EDITION)

E1   PROTECTION AGAINST SOUND FROM OTHER PARTS OF THE BUILDING AND ADJOINING BUILDINGS

Dwelling-houses, flats and rooms for residential purposes shall be designed and constructed in such a way that they provide reasonable resistance to sound from other parts of the same building and from adjoining buildings.

E2   PROTECTION AGAINST SOUND WITHIN A DWELLING-HOUSE ETC.

Dwelling-houses, flats and rooms for residential purposes shall be designed and constructed in such a way that: (a) internal walls between a bedroom or a room containing a water closet, and other rooms; and (b) internal floors provide reasonable resistance to sound.

E3   REVERBERATION IN THE COMMON INTERNAL PARTS OF BUILDINGS CONTAINING FLATS OR ROOMS FOR RESIDENTIAL PURPOSES

The common internal parts of buildings which contain flats or rooms for residential purposes shall be designed and constructed in such a way as to prevent more reverberation around the common parts than is reasonable.

E4   ACOUSTIC CONDITIONS IN SCHOOLS

(1) Each room or other space in a school building shall be designed and constructed in such a way that it has the acoustic conditions and the insulation against disturbance by noise appropriate to its intended use.

(2) For the purposes of this Part – 'school' has the same meaning as in section 4 of the Education Act 1996; and 'school building' means any building forming a school or part of a school.

F   VENTILATION (2006 EDITION)

F1   MEANS OF VENTILATION

There shall be adequate means of ventilation provided for people in the building.

G   HYGIENE (1992 EDITION, AS AMENDED)

G1   SANITARY CONVENIENCES AND WASHING FACILITIES

(1) Adequate sanitary conveniences shall be provided in rooms provided for that purpose, or in bathrooms. Any such room or bathroom shall be separated from places where food is prepared.

(2) Adequate washbasins shall be provided in: (a) rooms containing water closets; or (b) rooms or spaces adjacent to rooms containing water closets. Any such room or space shall be separated from places where food is prepared.

(3) There shall be a suitable installation for the provision of hot and cold water to washbasins provided in accordance with paragraph 2.

(4)   Sanitary conveniences and washbasins to which this paragraph applies shall be designed and installed so as to allow effective cleaning.

G2   BATHROOMS

A bathroom shall be provided containing either a fixed bath or shower bath, and there shall be a suitable installation for the provision of hot and cold water to the bath or shower bath.

G3   HOT WATER STORAGE

A hot water storage system that has a hot water storage vessel which does not incorporate a vent pipe to the atmosphere shall be installed by a person competent to do so, and there shall be precautions: (a) to prevent the temperatures of stored water at any time exceeding 100°C; and (b) to ensure that the hot water discharged from safety devices is safely conveyed to where it is visible but will not cause danger to persons in or about the building.

H   DRAINAGE AND WASTE DISPOSAL (2002 EDITION)

H1   FOUL WATER DRAINAGE

(1)   An adequate system of drainage shall be provided to carry foul water from appliances within the building to one of the following, listed in order of priority: (a) a public sewer; or, where that is not reasonably practicable, (b) a private sewer communicating with a public sewer; or, where that is not reasonably practicable, (c) either a septic tank which has an appropriate from of secondary treatment or another wastewater treatment system; or where that is not reasonably practicable (d) a cesspool.

(2)   In this Part 'foul water' means water which comprises or includes (a) waste from a sanitary convenience, bidet or appliance for washing receptacles for foul waste; or (b) water which has been used for food preparation, cooking or washing.

H2   WATER TREATMENT SYSTEMS AND CESSPOOLS

(1)   Any septic tank and its form of secondary treatment, other wastewater treatment system or cesspool, shall be so sited and constructed that: (a) it is not prejudicial to the health of any person; (b) it will not contaminate any watercourse, underground water or water supply; (c) there are adequate means of access for emptying and maintenance;

and (d) where, relevant, it will function to a sufficient standard for the protection of health in the vent of a power failure.

(2) Any septic tank, holding tank which is part of a wastewater treatment system or cesspool, shall be: (a) of adequate capacity; (b) so constructed that it is impermeable to liquids; and (c) adequately ventilated.

(3) Where a foul water drainage system from a building discharges to a septic tank, wastewater treatment system or cesspool, a durable notice shall be affixed in a suitable place in the building containing information on any continuing maintenance required to avoid risks to health.

H3   RAINWATER DRAINAGE

(1) Adequate provision shall be made for rainwater to be carried from the roof of the building.

(2) Paved areas around the building shall be so constructed as to be adequately drained.

(3) Rainwater from a system provided pursuant to sub-paragraphs (1) or (2) shall discharge to one of the following, listed in order of priority: (a) an adequate soakaway or some other adequate infiltration system; or, where that is not reasonably practicable, (b) a watercourse; or, where that is not reasonably practicable, (c) a sewer.

H4   BUILDING OVER SEWERS

(1) The erection or extension of a building or work involving the underpinning of a building shall be carried out in a way that is not detrimental to the building or building extension or to the continued maintenance of the drain, sewer or disposal main.

(2) In this paragraph 'disposal main' means any pipe, tunnel or conduit used for the conveyance of effluent to or from a sewage disposal works, which is not a public sewer.

(3) In this paragraph or paragraph H5 'map of sewers' means any records kept by a sewerage undertaker under section 199 of the Water Industry Act 1991.

H5   SEPARATE SYSTEMS OF DRAINAGE

Any system for discharging water to a sewer which is provided pursuant to paragraph H3 shall be separate from that provided for the conveyance of foul water from the building.

H6   SOLID WASTE STORAGE

    (1)   Adequate provision shall be made for storage of solid waste.

    (2)   Adequate means of access shall be provided: (a) for people in the building to the place of storage; and (b) from the place of storage to a collection point.

J   COMBUSTION APPLIANCES AND FUEL STORAGE SYSTEMS (2002 EDITION)

J1   AIR SUPPLY

Combustion appliances shall be so installed that there is an adequate supply of air to them for combustion, to prevent over-heating and for the efficient working of any flue.

J2   DISCHARGE OF PRODUCTS OF COMBUSTION

Combustion appliances shall have adequate provision for the discharge of products of combustion to the outside air.

J3   PROTECTION OF BUILDING

Combustion appliances and fluepipes shall be so installed, and fireplaces and chimneys shall be so constructed and installed, as to reduce to a reasonable level the risk of people suffering burns or the building catching fire in consequence of their use.

J4   PROVISION OF INFORMATION

Where a hearth, fireplace, flue or chimney is provided or extended, a durable notice containing information on the performance capacities of the hearth, fireplace, flue or chimney shall be affixed in a suitable place in the building for the purpose of safe installation of combustion appliances.

J5   PROTECTION OF LIQUID FUEL STORAGE SYSTEMS

Liquid fuel storage systems and the pipes connecting them to combustion appliances shall be so constructed and separated from buildings and the boundary of the premises as to reduce to a reasonable level the risk of the fuel igniting in the event of fire in adjacent buildings or premises.

J6   PROTECTION AGAINST POLLUTION

Oil storage tanks and pipes connecting them to combustion appliances shall: (a) be so constructed and protected as to reduce to a reasonable level

the risk of the oil escaping and causing pollution; and (b) have affixed in a prominent position a durable notice containing information or how to respond to an oil escape so as to reduce to a reasonable level the risk of pollution.

K   PROTECTION FROM FALLING, COLLISION AND IMPACT (1998 EDITION, AMENDED 2000)

K1   STAIRS, LADDERS AND RAMPS

Stairs, ladders and ramps shall be so designed, constructed and installed as to be safe for people moving between different levels in or about the building.

K2   PROTECTION FROM FALLING

(a) Any stairs, ramps, floors and balconies and any roof to which people have access, and (b) any light well, basement area or similar sunken area connected to a building, shall be provided with barriers where it is necessary to protect people in or about the building from falling.

K3   VEHICLE BARRIERS

(1)   Vehicle ramps and any levels in a building to which vehicles have access, shall be provided with barriers where it is necessary to protect people in or about the building.

(2)   Vehicle loading bays shall be constructed in such a way, or be provided with such features, as may be necessary to protect people in them from collision with vehicles.

K4   PROTECTION FROM COLLISION WITH OPEN WINDOWS ETC.

Provision shall be made to prevent people moving in or about the building from colliding with open windows, skylights or ventilators.

K5   PROTECTION AGAINST IMPACT FROM AND TRAPPING BY DOORS

(1)   Provision shall be made to prevent any door or gate: (a) which slides or opens upwards, from falling onto any person: and (b) which is powered, from trapping any person.

(2)   Provision shall be made for powered doors and gates to be opened in the event of a power failure.

(3)   Provision shall be made to ensure a clear view of the space on either side of a swing door or gate.

L CONSERVATION OF FUEL AND POWER (2006 EDITION)

L1A NEW DWELLINGS

Reasonable provision shall be made for the conservation of fuel and power in buildings by: (a) limiting heat gains and losses (i) through thermal elements and other parts of the building fabric; (ii) from pipes, ducts and vessels used for space heating, space cooling and hot water services; (b) providing and commissioning energy efficient fixed building services with effective controls; and (c) providing to the owner sufficient information about the building, the fixed building services and their maintenance requirements so that the building can be operated in such a manner as to use no more fuel and power than is reasonable in the circumstances.

L1B EXISTING DWELLINGS

As L1A.

L2A NEW BUILDINGS OTHER THAN DWELLINGS

As L1A.

L2B EXISTING BUILDINGS OTHER THAN DWELLINGS

As L1A.

M ACCESS TO AND USE OF BUILDINGS (2004 EDITION)

M1 ACCESS AND USE

Reasonable provision shall be made for people to: (a) gain access; and (b) use the building and its facilities.

M2 ACCESS TO EXTENSIONS TO BUILDINGS OTHER THAN DWELLINGS

Suitable independent access shall be provided to the extension where reasonably practicable.

M3 SANITARY CONVENIENCES IN EXTENSIONS TO BUILDINGS OTHER THAN DWELLINGS

If sanitary conveniences are provided in any buildings that is to be extended, reasonable provision shall be made within the extension for sanitary conveniences.

M4   SANITARY CONVENIENCES IN DWELLINGS

(1)   Reasonable provision shall be made in the entrance storey for sanitary conveniences, or where the entrance storey contains no habitable rooms, reasonable provision for sanitary conveniences shall be made in either the entrance storey or principal storey.

(2)   In this paragraph 'entrance storey' means the storey which contains the principal entrance and 'principal storey' means the storey nearest to the entrance storey which contains a habitable room, or if there are two such storeys equally near, either such storey.

N   GLAZING – SAFETY IN RELATION TO IMPACT, OPENING AND CLEANING (1998 EDITION, AMENDED 2000)

N1   PROTECTION AGAINST IMPACT

Glazing, with which people are likely to come into contact while in passage in or about the building, shall: (a) if broken on impact, break in a way which is unlikely to cause injury; or (b) resist impact without breaking; or (c) be shielded or protected from impact.

N2   MANIFESTATION OF GLAZING

Transparent glazing, with which people are likely to collide while in passage in or about the building, shall incorporate features which make it apparent.

N3   SAFE OPENING AND CLOSING OF WINDOWS ETC.

Windows, skylights and ventilators which can be opened by many people in and about the building shall so be constructed or equipped that they may be opened, closed or adjusted safely.

N4   SAFE ACCESS FOR CLEANING WINDOWS ETC.

Provision shall be made for any windows, skylights or any transparent or translucent walls, ceilings or roofs to be safely accessible for cleaning.

P   ELECTRICAL SAFETY (2006 EDITION)

P1   DESIGN AND INSTALLATION

Reasonable provision shall be made in the design and installation of electrical installations in order to protect persons operating, maintaining or altering the installations from fire or injury.

Appendix B
Hazard Identification Checklist

Identified hazards form the basis for risk assessment, which assesses the likelihood of harm being caused. Once this is known the risk may be managed by either removing the hazard or reducing the likelihood or severity of resulting harm by re-planning the work process or activity. A 'safe system of work' or 'method statement' may be required to document how the activity will be carried out safely.

Hazards should be identified before visiting premises/sites, reviewed upon arrival and during survey or inspection, and risks managed accordingly.

This checklist is based on *Surveying Safely: Your Guide to Personal Safety at Work* (RICS, 2004), and is reproduced by permission of the Royal Institution of Chartered Surveyors which owns the copyright.

BEFORE VISITING PREMISES/SITES

Travelling to and from site

- Plan the journey to avoid driving too fast, for too long or when tired
- Be aware of where to park (clear, secure, easy to exit, well lit)

Lone working

- Provision for communications in an emergency
- Record of where the lone worker is and when expected back in office or at home
- Procedures for regular 'check-in' calls
- Access for rescue
- Medical condition of lone worker (e.g. epilepsy, diabetes)

Condition of site

- State of construction, site rules
- Derelict premises, poor condition, nature of damage
- Areas deemed unsafe for access
- Security measures, access arrangements
- Need for protective clothing or special equipment

Occupation

- Occupied premises, prior arrangements, special access provision

- Who will be encountered in building or on site (e.g. children, squatters, vagrants, animals)
- Aggressive or disaffected occupants or neighbours

Activity
- Nature of occupation (e.g. residential, manufacturing, warehousing) and what might be encountered (e.g. noise, fumes, vehicle movements, electronic equipment)

Site rules and welfare
- Specific 'house rules' of client or property manager
- 'Permit to work/enter' rules
- Availability of 'construction phase health and safety plan' for construction sites including induction procedures
- Availability of toilet, wash, and first aid facilities

High structures
- Is scaffolding safe to use, when was it last inspected
- Access to towers, masts or chimneys
- Provision and operation of special access equipment (e.g. cherry picker)

Dangerous substances
- Potential contact with hazardous substances (e.g. chemicals, radiation, asbestos, gas, explosives)
- Availability of registers or records (e.g. asbestos, environmental hazards), special precautions

Diseases
- Contamination with any form of clinical waste
- Used syringes/needles, condoms, razor blades
- Potential source of anthrax
- Potential source of *Legionella* (e.g. disused water storage systems)
- Hazards from vermin (e.g. Weil's disease)

Special access
- Provision and management of special access arrangements
- Requirements for special training

Special risks
- Building or sites with special hazards (e.g. railway premises, security establishments, confined spaces, plant rooms)

Special equipment
- Building or site requiring special equipment (e.g. gloves, respirator or face mark, safety helmet, ear defenders, eye protection, boots, temporary lighting)

Environmental
- Weather conditions
- Light levels
- Temperature extremes

Personal
- Issues of gender, pregnant or nursing mothers, levels or lack of fitness
- Requirement for special skills

- Phobias, vertigo or claustrophobia that would impair judgement regarding personal safety

ARRIVING AND DURING VISITS TO PREMISES/SITES

Structures

The chance of partial of total collapse of:

- Chimney stacks, gable walls or parapets
- Leaning, bulged and unrestrained walls (including boundary walls)
- Rotten or corroded beams and columns
- Roofs and floors

Timbers and glass

- Rotten and broken floors and staircases, flimsy cellar flaps and broken pavement lights
- Floorboards, joists and buried timbers weakened by age, decay or attack
- Projecting nails and screws, broken glass
- Glazing in windows and partitions may be loose, hinges and sashcords weak or broken. Glass panels in doors and winglights may be painted over

Roofs

- Fragile asbestos cement and plastic coverings
- Fragile rooflights (often obscured by dirt or temporary coverings)

- Low parapets or unguarded roof edges, loose copings
- Rusted, rotten or moss covered fire escapes, access ladders and guard rails
- Rotten roof decking and joists
- Slippery roof coverings (slates, moss or algae covered slopes)
- Broken access hatches
- Mineral wool dust, mortar droppings and birds' nesting material and excrement in roof voids Cornered birds and vermin
- Insects, including bee and wasp colonies
- Water-cooling plant may harbour *Legionella*
- Unguarded flat roofs
- Broken, loose, rotten and slippery crawling boards and escape ladders
- Weak flat roofs and dust covered rooflights
- Slippery roof surfaces
- High winds during roof inspection
- Ill-secured or flimsy, collapsible, sectional or fixed loft ladders
- Concealed ceiling joists and low purlins

Ill-lit roof voids

- Unsafe atmosphere
- Confined spaces with insufficient oxygen including manholes, roof voids, cellars, vaults, ducts and sealed rooms
- Rotting vegetation which may consume oxygen and give off poisonous fumes
- Accumulation of poisonous or flammable gases in buildings on contaminated land

- Stores containing flammable materials such as paint, adhesives, fuel and cleaning fluids
- Hazardous substances, including toxic insecticides and fungicides
- Gas build-up in subfloor voids

Danger from live and unsecured services

- Electricity, gas, water and steam supplies
- Awkward entrances into substations and fuel stores
- Temporary lighting installations: mains connections and generators
- Buried cables and pipes
- Overhead electrical cables

Hidden traps, ducts and openings

- Lift and services shafts, stairwells and other unguarded openings
- Manholes, including those obscured by flimsy coverings. Cesspools, wells and septic tanks

Intruders and others

- Physical dangers from squatters, vagrants or guard dogs
- Disease risks from discarded syringes and condoms
- Structures weakened by vandalism or arson
- Aggressive tenants and property owners

Contamination

- Asbestos, lead and other substances hazardous to health

- Chemicals in storage or leaked
- Contaminated water supplies
- Contaminated air conditioning systems (*Legionella*)

Vermin and birds

- Rats and mice: Weil's and other diseases
- Bird droppings
- Lice and fleas may be present in bedding, soft furniture and carpets

WHEN VISITING
CONSTRUCTION SITES

- Slips, trips and falls
- Unsafe scaffolding and ladders
- Deep unsafe or unsupported excavations
- Cranes and overhead hazards
- Uneven ground
- Discarded materials, especially those with projecting nails
- Electricity
- Unsighted and reversing vehicles
- Inspecting highways or vehicle access points without illuminated clothing and hazard signs
- Hot bitumen and asphalt
- Wet surfaces

ON MINING AND SIMILAR SITES

Land and property damaged by mining subsidence

- Uneven ground surface and paved areas
- Loose or structurally unsound walls, floors, roofs, fixtures and fittings

- Gas leaks
- Fissuring

Mine shafts and adits and shallow mine workings
- Unstable ground around the shaft/adit
- Possible existence of nearby shafts with disturbed cappings and filling which may have a potential for collapse
- Toxic and explosive gases emanating from the shaft or adit
- Pitfalls and crown holes associated with old shallow underground mining activities
- Danger from the mining operation and machinery

Quarries
- Pitfalls and shallow mining
- Steep and/or unstable quarry faces and benches
- Unstable ground at the top of quarry faces or benches
- Danger of loose material falling from above
- Moving parts of machinery
- Unguarded electrical and compressed air equipment
- Blasting
- Mobile excavations and large earth transporters
- Elevated walkways, stagings, platforms and ladders
- Hot surfaces at coating plants, etc.
- Slurry ponds, lagoons, tanks and other filled areas
- Noisy and dusty conditions
- Railways and internal haul roads

- Danger from the quarrying operation and mobile plant

Tips and land reclamation sites
- Unstable slopes and ground
- Water lagoons, ponds and other water filled areas
- Slurry and quicksand areas
- Burning areas where tips are heating or on fire
- Hazardous or harmful chemicals, liquid matters and wastes, contaminated land
- Explosive and toxic gases and vapours

Gas and oil wells
- Pipes and other ground-level hazards
- Flare stacks
- Separating lagoons
- Explosive atmospheres
- Danger from the drilling operations

Exploration, drilling and gantry sites
- Hot muds
- Flying rock, dust and debris
- Water hazards
- Unsafe plant

ON FARMS

Farm buildings and land
- Grain storage and handling installations, particularly moving augers and conveyors
- Underground slurry stores, slurry lagoons, drains, deep ditches, wells, tower silos, sewage tanks and silage clamps (note: risk of toxic gases)

- Dust hazards in grain, mill and mix and intensive livestock buildings
- Overgrown areas: concealed manholes
- Poorly maintained buildings – especially loft floors – and fittings
- High-voltage electric fencing
- Stored hazardous chemicals
- Rivers, lakes, reservoirs, dangerous bridges, bogs, quicksands, unstable cliff edges and the sea
- Chemicals, poisons

Farm machinery
- Packing and grading machines
- Stones and debris thrown from swipes and hedgecutters
- Cranes and lifting equipment

Livestock
- Any entire male animal: bulls, boars and rams
- Any female animal with young – calf-proud cows and farrowing sows
- Game parks and wild animals
- Horses
- Dogs

Diseases and pests
- Tetanus, brucellosis
- Weil's disease (from stagnant water and ponds, hay stress, etc.)

Sporting
- Firearms (must be licensed and properly stored and used)

IN FORESTS

- Tree-felling work, either in thinning or clear felling operations
- Tree surgery work
- Dangerous and damaged trees, especially if liable to shed limbs
- Any work in woodlands in high winds
- Hand tools, axes and swipes
- Chainsaws
- Saw milling and cutting equipment in saw mills and wood yards
- Timber handling equipment (e.g. overhead extraction lines)
- Fork-lift vehicles and cranes

OFFSHORE

- Moving and often slippery decks of ships and oil rigs
- Poorly secured equipment, including coffee cups and tools near sensitive machinery
- Fire hazards and flammable materials
- The hostile deck environment
- High pressure air gases and gas storage cylinders
- Hydraulic oil leakage

Appendix C
Useful Contacts

Aluminium Federation
www.alfed.org.uk
Tel: 0121 456 1103

Arboricultural Association
www.trees.org.uk
Tel: 01794 368717

Association d'Experts Européens du Bâtiment et de la Construction
www.aeebs.org

Association for Environment Conscious Building (AECB)
Tel: 0845 4569773
www.aecb.net

Avoncroft Museum of Historic Buildings
www.avoncroft.org.uk

Avongard Limited
Tel: 01275 849782
www.avongard.co.uk

Barn Owl Trust
Tel: 01364 653026
www.barnowltrust.org.uk

Bat Conservation Trust
Tel: 020 7627 2629
www.bats.org.uk

Brick Development Association (BDA)
Tel: 01344 885651
www.brick.org.uk

British Geological Survey
Tel: 0115 936 3100
www.bgs.ac.uk

British Institute of Non-Destructive Testing
Tel: 01604 630124
www.bindt.org

British Lime Association
Tel: 020 7963 8000
www.britishlime.org

British Plastics Federation
Tel: 020 7457 5000
www.bpf.co.uk

British Standards Association (BSI)
Tel: 020 8996 9001
www.bsi-global.com

British Wood Preserving and Damp-Proofing Association
Tel: 01332 225100
www.bwpda.co.uk

Brooking Collection
Tel: 020 8331 9897
www.gre.co.uk

Building Centre
Tel: 020 7692 4000
www.buildingcentre.co.uk

Building Limes Forum
www.builidinglimesforum.org.uk

Building Performance Group
Tel: 020 7583 9502
www.bpg-uk.com

Building Research Establishment (BRE)
Tel: 01923 664664
www.bre.co.uk

Cambridge University Collection of Aerial Photographs
Tel: 01223 334575
www.aerial.cam.ac.uk

Centre for Accessible Environments
Tel: 020 7840 0125
www.cae.org.uk

Centre for Sustainable Construction
Tel: 01923 664500
www.bre.co.uk

Chartered Institute of Building (CIOB)
Tel: 01344 630700
www.ciob.org.uk

Chartered Institution of Building Services Engineers (CIBSE)
Tel: 020 8675 5211
www.cibse.org.uk

Chiltern Open Air Museum
Tel: 01494 871117
www.coam.org.uk

Common Ground
Tel: 01747 850820
www.commonground.org.uk

Construction History Society
Tel: 01344 630741
www.constructionhistory.co.uk

Construction Resources
Tel: 020 7450 2211
www.constructionresources.com

Copper Development Association (CDA)
Tel: 01442 275 700
www.cda.org.uk

Ecology Building Society
Tel: 0845 674 5566
www.ecology.co.uk

Empty Homes Agency
Tel: 020 7828 6288
www.empty.homes.com

English Heritage
Tel: 0870 333 1187
www.english-heritage.org.uk

Fire Protection Association (FPA)
Tel: 01608 872500
www.thefpa.co.uk

Forest Stewardship Council (FSC)
Tel: 01686 413916
www.fsc-uk.info

Glass and Glazing Federation
www.ggf.org.uk

Grant Instruments (Cambridge)
Limited
Tel: 01763 260811
www.grant-dataacquisition.com

Green Register
Tel: 0117 377 3490
www.greenregister.org

Gypsum Products Development
Association
Tel: 020 7935 8532
www.gpda.org.uk

Hanwell Instruments Limited
Tel: 0870 443 1786
www.hanwell.com

Health and Safety Executive (HSE)
Tel: 0845 345 0055
www.hse.gov.uk

Health Protection Agency (HPA)
Tel: 020 7759 2700
www.hpa.org.uk

Housing Association Property
Mutual (HAPM)
Tel: 020 7204 2424
www.buildinglifeplans.com

Hutton + Rostron Environmental
Investigations Limited
01483 203221
www.handr.co.uk

International Council for Research
and Innovation in Building
Construction (CIB)
Tel: 00 31 10 411 0240
www.cibworld.nl/website

International Council on
Monuments and Sites (ICOMOS)
Tel: +33(0)1 45 67 67 70
www.icomos.org

Ironbridge Gorge Museums
Tel: 01952 884391
www.ironbridge.org.uk

Landmark Information Group
Limited
Tel: 01392 441700
www.landmarkinfo.co.uk

Lead Sheet Association (LSA)
Tel: 01892 822773
www.leadsheetassociation.org.uk

Mastic Asphalt Council and
Employers Federation
Tel: 01424 814400
www.masticasphaltcouncil.co.uk

Men of the Stones
Tel: 01952 850269
www.menofthestones.org.uk

Meteorological Office
Tel: 0870 900 0100
www.met-office.gov.uk

Museums, Libraries and Archives
Council (MLA)
Tel: 020 7273 1444
www.mla.gov.uk

National Monuments Record
Centre (NMR)
Tel: 01793 414600
www.english-heritage.org.uk

National Monuments Record
of Wales
Tel: 01970 621200
www.rcahmw.gov.uk

National Monuments Record
of Scotland
Tel: 0131 662 1456
www.rcahms.gov.uk

National Trust
Tel: 01793 817400
www.nationaltrust.org.uk

Natural Building Technologies
Limited
Tel: 01844 338338
www.natural-building.co.uk

Natural England
Tel: 0845 600 3078
www.naturalengland.gov.uk

Paint Research Association
Tel: 020 8487 0800
www.pra-world.com

Practical Action
Tel: 01926 634400
www.practicalaction.org.uk

Preservation Equipment Limited
Tel: 01379 647400
www.preservationequipment.co.uk

Royal Institute of British
Architects (RIBA)
Tel: 020 7580 5533
www.riba.org.uk

Royal Institution of Chartered
Surveyors (RICS)
Tel: 0870 333 1600
www.rics.org

Royal Town Planning Institute
(RTPI)
Tel: 020 7929 9494
www.rtpi.org.uk

St Fagans National History
Museum
Tel: 029 2057 3500
www.museumwales.ac.uk

Salvo
Tel: 020 8400 6222
www.salvo.co.uk

Society for the Protection of
Ancient Buildings (SPAB)
Tel: 020 7377 1644
www.spab.org.uk

Stainless Steel Advisory Centre
Tel: 0114 267 1265
www.bssa.org.uk

Stone Federation of Great Britain
Tel: 01303 856123
www.stone-federation.org.uk

Stone Roofing Association
Tel: 01286 650402
www.stoneroof.org.uk

Sustainability Centre
Tel: 01730 823166
www.earthworks-trust.com

Thatching Advisory Services
Limited
Tel: 01264 773820
www.thatchingadvisoryservices.
co.uk

Tiles and Architectural Ceramics
Society
www.tilesoc.org.uk

Timber Research and Development
Association (TRADA)
Tel: 01494 569600
www.trada.co.uk

Traditional Paint Forum
www.traditionalpaintforum.org.uk

Upkeep (The Trust for Training
and Education in Building Repairs
and Maintenance)
Tel: 020 7631 1677
www.upkeep.org.uk

Vincent Wildlife Trust
Tel: 01531 636441
www.vwt.org.uk

Weald and Downland Open Air
Museum
Tel: 01243 811363
www.wealddown.co.uk

Zinc Information Centre
Tel: 0121 362 1201
www.zincinfocentre.org

Glossary

Absorption   Penetration of one substance, such as water, into the body of another.

Adsorption   Formation of a layer of one substance on the surface of another.

Amorphous   Non-crystalline solid (such as glass).

Anthropodynamic   Interface between the occupants and the building; aspects of design related to dexterity and manoeuvrability.

Anthropogenic   Produced or caused by humans.

Building pathology   Identification, investigation and diagnosis of defects in existing buildings; prognosis of defects and recommendations for the most appropriate course of action having regard to the building, its future and resources available; and design, specification, implementation and supervision of appropriate programmes of remedial works, with monitoring and evaluation in terms of functional, technical and economic performance in use.

Capillarity   Capacity of a liquid to move upwards or downwards within the fine pore spaces of a material due to the effects of surface tension.

Condensation   Process of forming a liquid from its vapour.

Cost–benefit analysis   Assessment of the desirability of projects, where the indirect effects on third parties outside those affecting the decision-making process are taken into account.

Crypto-efflorescence   Deposition of soluble salts beneath the surface of a porous material as a result of the evaporation of water in which the salts are dissolved.

Crystalline   Having a regular internal arrangement of atoms, ions or molecules.

Defect   Non-fulfilment of an intended requirement or an expectation, including that concerned with safety.

Deformation   Change in shape of a material due to the application or inducement of a force.

Deliquescent   Absorption of water from the atmosphere by a hygroscopic solid to such an extent that a concentrated solution of the solid eventually forms.

Density   Mass of substance per unit volume.

Dew-point   Temperature at which air would become saturated if cooled at constant pressure.

Diagnosis   Deciding the nature of a fault from its symptoms.

Ductility   Capacity of a material, usually a metal, to be drawn out plastically before breaking.

Durability   Ability of a building and its parts to perform its required functions over a period of time and under the influence of internal and external agencies or mechanisms of deterioration and decay.

Efflorescence   Deposition of soluble salts at the surface of a porous material as a result of the evaporation of water in which the salts are dissolved.

Elasticity   Property of a material that enables it to return to its original shape and form once the stress causing the deformation has been removed.

Electrochemical reaction   Reaction involving ions in solution.

Equilibrium moisture content   Moisture content of a material that it will achieve when it is in equilibrium with the moisture content of the surrounding air.

Evaporation   Process whereby the quantity of a liquid exposed to air is progressively reduced until it eventually disappears.

Facilities management   Practice of coordinating the physical workplace with people and work of the organisation; integrates the principles of business administration, architecture and the behavioural and engineering sciences.

Fatigue   Fracture of materials when subjected to fluctuating or repeated load that is within the stress limit for static loading.

Fault   State characterised by an inability to perform a required function, excluding the ability during preventive maintenance or other planned actions, or due to a lack of external resources.

Feng shui   System of organising the home and workplace in a way that promotes health, happiness and success; art of building design that is solely focused on the success of the occupants.

Force   Physical agent that causes a change in momentum or elastic strain in a body.

Gas   State of matter having no definite volume or shape, but filling any vessel into which it is put.

Hydrophobic   Lacking affinity for water.

Hygroscopic   Describing a substance that can take up water from the atmosphere.

Information management   An organised and structured approach to handling information and data so as to ensure that the right information is provided to the right people at the right time and in the right format.

Life-cycle   Time interval that commences with the initiation of a concept and terminates with the disposal of the asset.

Life-cycle costs   Total cost of ownership of an item, taking into account all the costs of acquisition, personnel training, operation, maintenance, modification and disposal, for the purpose of making decisions on new or changed requirements and as a control mechanism in service for existing and future items.

Liquid   State of matter having a definite volume, but no definite shape (taking its shape from that of the containing vessel).

Maintainability   Probability that a given maintenance action under given conditions of use can be carried out within a stated time interval, when the maintenance is performed under stated conditions and using stated procedures and resources.

Maintenance management   Structured planning, control and implementation of maintenance activities.

Malleability   Ability of a material, usually a metal, to be beaten into sheets without rupturing.

Metastable   Condition of a system in which it has a precarious stability that can easily be disturbed.

Microclimate   Climate in a very small area or in a particular location with specific conditions that is compared to the general climate of which it is a part.

Moisture content   Amount of moisture that a material contains at a given time, expressed as a percentage of its dry mass.

Permeability   Extent to which a material will allow a substance to pass through it.

pH   Logarithmic scale for expressing the acidity or alkalinity of a solution based on the concentration of hydrogen ions; a neutral solution has a pH of 7, whilst a pH below 7 indicates an acid solution and above 7 indicates an alkaline solution.

Photochemical reaction   Reaction caused by light or ultraviolet radiation.

Physicochemical reaction   Reaction involving both physics and chemistry.

Planned maintenance   Maintenance organised and carried out with forethought, control and the use of records, to a predetermined plan based on the results of previous condition surveys.

Pores   Spaces between the particles of which a material is composed.

Porosity   Ratio of volume of voids to that of the overall volume of a material.

Preventive maintenance   Maintenance carried out at predetermined intervals, or corresponding to prescribed criteria, and intended to reduce the probability of failure or performance degradation of an item.

Prognosis   Predicting or forecasting the course of a fault from its symptoms.

Project management   Overall planning, control and coordination of a project from inception to completion, aimed at meeting a client's requirements and ensuring completion on time, within cost and to required quality standards.

Salts   Formed as a result of a chemical reaction such as between an acid and an alkali, the most common being carbonates, chlorides, nitrates and sulphates.

Solid   State of matter, whether crystalline or amorphous (non-crystalline), having a definite volume and shape.

Solvent   Liquid that holds a solid in solution.

Strain   Measure of deformation produced by an acting force, relating change in form with that of the original form prior to loading.

Strength   Ability of a material to sustain loads without undue distortion or fatigue.

Stress   Intensity of internal forces mobilised to resist deformation caused by external force.

Symbiosis   Interaction between individuals of different species; usually denoting to interactions in which both species benefit.

Terotechnology   Combination of management, financial, engineering, building and other practices applied to physical assets in the pursuit of economic life-cycle costs; concerned with the specification and design of reliability and maintainability of physical assets such as plant, machines, equipment, buildings and structures.

Whole life-cycle cost   Generic term for the costs associated with owning and operating a facility from inception to demolition, including both initial capital costs and running costs.

Index

Printed and bound by CPI Group (UK) Ltd, Croydon, CR0 4YY